GALAXIES

—✳—

Galaxies

INSIDE THE UNIVERSE'S
STAR CITIES

DAVID J. EICHER

Illustrated by Irene Laschi

CLARKSON POTTER / PUBLISHERS
New York

Copyright © 2020 by David Eicher
Illustrations copyright © 2020 by Irene Laschi

Published in the United States by Clarkson Potter/
Publishers, an imprint of Random House, a division of
Penguin Random House LLC, New York.
clarksonpotter.com

CLARKSON POTTER is a trademark and POTTER
with colophon is a registered trademark of Penguin
Random House LLC.

Library of Congress Cataloging-in-Publication Data
Names: Eicher, David J., 1961– author. | Laschi, Irene,
    1984– illustrator.
Title: Galaxies : inside the universe's star cities / David
    J. Eicher ; illustrations by Irene Laschi.
Description: New York : Clarkson Potter/Publishers,
    [2020] | Includes bibliographical references
    and index. | Summary: "Tour the most dazzling,
    fascinating, and unusual galaxies in the universe
    with the Editor-in-Chief of Astronomy as your
    personal guide, featuring jaw-dropping illustrations
    and full-color photography from the magazine's
    archives, much of it never before published."—
    Provided by publisher.
Identifiers: LCCN 2019028797 (print) | LCCN
    2019028798 (ebook) | ISBN 9780525574316
    (hardback) | ISBN 9780525574323 (ebk)
Subjects: LCSH: Galaxies.
Classification: LCC QB857 .E33 2020 (print) | LCC
    QB857 (ebook) | DDC 523.1/12—dc23
LC record available at https://lccn.loc.gov/
    2019028797
LC ebook record available at https://lccn.loc.gov/
    2019028798

ISBN 978-0-525-57431-6
Ebook ISBN 978-0-525-57432-3

Printed in China

Book and cover design by Mia Johnson
Illustrations by Irene Laschi

10 9 8 7 6 5 4 3 2 1

First Edition

Previous **M106: A SWIRLING DISK OF GAS AND STARS**
The spiral galaxy M106 in Canes Venatici lies at a distance of
24 million light-years and is bright enough to be a familiar
target for galaxy observers armed with telescopes. It was
one of the first galaxies with a recognized supermassive
central black hole. In fact, this is a Seyfert galaxy, with an
irregularly active energy output, and its black hole has
about the same mass of the Milky Way's—3.9 million solar
masses. M106 also contains Cepheid variable stars that
are key to helping set the cosmic distance scale, allowing
astronomers to refine their distance measurements.

THE MAGELLANIC CLOUDS OVER THE ATACAMA DESERT
Countless stars from our galaxy, the Milky Way, appear in
this image taken in Chile's high Atacama Desert, perhaps the
darkest sky on Earth. The Large and Small Magellanic Clouds,
satellite galaxies of the Milky Way, stand out in the sky.

Overleaf **THE BLACK EYE GALAXY IN COMA BERENICES**
Another showpiece target for amateur skywatchers, the
Black Eye Galaxy (M64) in Coma Berenices features a broad,
arcing dust lane across the galaxy's hub that gives the
object its nickname. This galaxy consists of counter-rotating
disks, which may have arisen from an earlier collision. It lies
17 million light-years away.

FOR BRIAN MAY,
who makes the galaxy a far more
interesting place in which to live.

CONTENTS

Foreword

✶

n the first quarter of the twentieth century the existence of galaxies moved from a matter of speculation to observational fact. By the early 1930s distances (albeit inaccurate due to problems that were not recognized at the time) were estimated to several nearby galaxies and the expansion of the universe was established. The Milky Way became part of a vast population of galaxies that are illuminated by their stellar populations. In optical images galaxies appeared to be well separated in space, and the concept of galaxies as isolated regions, "island universes," was born, a view that remained in place for decades.

After 1950, with the advent of large optical telescopes and new technologies, our model for galaxies grew richer. Radio observations revealed the presence of often billions of solar masses of interstellar gas, as well as showing that some galaxies emitted much of their power at radio wavelengths, a process requiring massive injections of electrons moving at nearly the speed of light. These data and X-ray observations led to the recognition that multimillion solar mass black holes lurk in the centers of most giant galaxies.

With the confirmation of the Big Bang model, it was obvious that galaxies had to form in the early universe and then evolve into their present-day structures. However, theoretical models for galaxies consisting only of normal baryonic matter failed to produce credible results. The answer came from an unexpected quarter. In disk galaxies the rotation speeds at large radii failed to decline as required if

8

galaxies consisted only of visible stars and gas. An extra invisible component, the dark matter halos, evidently surround galaxies and make up most of their mass. Dark matter dominated models of galaxies not only fit the data but also allowed for the development of a testable theoretical framework for galaxy formation and subsequent evolution.

In the late twentieth century the Hubble Space Telescope and ground-based 8- to 10-m aperture telescopes opened the way to the study of youthful galaxies. One of astronomy's great advantages is the ability to directly see the distant past, provided, of course, that one has sufficient sensitivity! These studies and their multi-wavelength follow-ons revealed that young galaxies are compact, interactive, and potentially supporting stupendous star formation rates. In other cases power outputs equivalent to trillions of suns were supplied by young supermassive black holes, as they rapidly grew by accreting gas and likely also stars. Consistent with the hierarchical growth through mergers concept in dark matter models, galaxies in their youth were anything but the stately island universes of the 1930s!

So where are we now? We know that galaxies are complex physical systems that experience dramatic evolution as they age, interact with their environments as they convert gas into long-term storage in stars or sequester normal matter into black holes. The presence of dark matter halos is the standard model, one that we constantly check through comparisons between ever improving simulations of the development of cosmic structure in an evolving universe and increasingly rich observations. With the power provided by high performance telescopes operating across much of the electromagnetic spectrum, we now can admire and study the complexity of galaxies in ways that our eyes cannot see, and at the same time enjoy their celestial beauty when resolved into myriads of stars or fantastic structures of interstellar gas.

This book introduces the world of galaxies from a twenty-first-century perspective. We have learned a great deal since the existence of galaxies began to be seriously recognized almost exactly 100 years ago, but still have a long journey before we fully understand these natural marvels. Enjoy this guide with its many images of galaxies, and if you have a chance to observe a galaxy through a telescope, most likely as a smudge of light, you will more fully realize why what you are seeing is one of nature's wonders.

—JAY GALLAGHER
Madison, Wisconsin

VIBRANT COLORS OF THE ANDROMEDA GALAXY
This spectacular portrait of the Andromeda Galaxy
shows it in its full vibrant colors, with bluish spiral
arms filled with bursts of star formation, and an
older, yellowish population of stars near the
galaxy's central hub.

THE MILKY WAY'S COMING COLLISION WITH ANDROMEDA
This image shows one moment in the future collision of the Milky Way with the Andromeda Galaxy, which will produce a chaotic supergalaxy, a merger some 4 to 5 billion years from now.

Introduction

---✳---

GAZING INTO A SUMMER SKY

When I was fourteen years old, I was invited to a "star party" in the little town of Oxford, Ohio, where I grew up. Someone had a six-inch telescope there, and I was immediately entranced by the realization that I could walk outside into a yard and gaze deeply into the universe. It was Saturn that caught my attention at that first star party, but soon I was out in a large cornfield in the back of my own subdivision, night after night, exploring the sky with a simple pair of 7x50 binoculars.

What a spectacular summer of binocular stargazing that was! I knew almost nothing of the sky, and had no telescope. Every new view was a discovery, whether it was scanning along the glistening backdrop of the Milky Way, spotting a star cluster, or coming across a bright and colorful star. I took the first steps in a deep knowledge of the sky that summer, something that too few stargazers now possess, in an era of computerized go-to telescopes.

Soon I wandered my binoculars' field of view into the Great Square of Pegasus, and across the nearby constellation Andromeda. The glow I encountered there, like a bright hazy star surrounded by an elongated, oval fuzz, I soon learned was a pretty special "deep-sky object."

I had stumbled onto the Andromeda Galaxy, and as soon as I learned what it was, I also learned to see it with my eye alone under our dark Ohio sky. In fact, for most people, it is the most distant object visible to the unaided human eyes. The Andromeda Galaxy is a galaxy like our Milky Way, located some 2.5 million light-years away—that's 57 million trillion kilometers, a long hike. (Under perfect sky conditions, some experienced observers claim to see more distant galaxies like M33 and M81 with the naked eye.)

Until the early 1920s, no one knew what galaxies were. Moreover, they had no idea how large the universe was. For many decades beforehand, "spiral nebulae" were thought to be strange gas clouds within our own galaxy. In the early 1920s American astronomer Edwin Hubble made a breakthrough discovery at Mount Wilson Observatory that led to understanding the fundamental nature of galaxies. Aided by work by Lowell Observatory astronomer Vesto M. Slipher, Hubble and others deciphered the cosmic distance scale, at least to a first approximation. By the late 1920s astronomers and the informed public understood that we live in a universe filled with galaxies and that the Milky Way and the Andromeda Galaxy are just two of them.

As I learned more and more about galaxies, I turned to every book I could get my hands on. I was immersed in these distant objects right away. A small astronomy club called the Astronomical Association of Oxford Ohio was built around chiefly college students in our community, which hosted Miami University in southwestern Ohio. They were looking for a columnist to write about galaxies and other distant objects in the sky, collectively known as deep-sky objects, the stuff beyond our solar system. I started writing a column about my observations, and this led to me starting a little "magazine"—a newsletter, really—produced at first on a mimeograph machine in my father's chemistry office at the university. *Deep Sky Monthly* reported on numerous galaxies, and grew over its five-year lifespan to a circulation of 1,000.

This growing interest in observing galaxies coincided with a burgeoning availability of larger telescopes in the hands of amateur astronomers, in what became known as the Dobsonian revolution. John Dobson, a San Francisco area amateur astronomer, pioneered a technique of making large telescopes with inexpensive mirrors, perched on simple mounts that moved up and down and sideways like a battleship gun turret. Larger telescopes in the hands of amateur astronomers meant that more and more people could see fainter and fainter objects for themselves, including numerous galaxies. Interest in observing the sky skyrocketed.

One of the books I discovered early on was *Galaxies*, a pictorial book with a smartly written scientific text by the great science writer Timothy Ferris. Published in 1980, the book contained numerous beautiful photographs and clever diagrams that showed the structure of our Milky Way galaxy and the universe of galaxies around us in a pseudo-3-D style. I loved this book and it was a big influence on me and my early astronomical interests.

In 1982, fresh from Miami University, I was hired as an assistant editor of *Astronomy* magazine, the world's largest publication on the subject, and went to Milwaukee, bringing my small magazine with me. Retitled *Deep Sky* and now a quarterly, the journal of observing galaxies, clusters, and nebulae reached a peak circulation of 15,000 during its ten-year lifetime. By 1992, our company decided that I shouldn't spend a quarter of my time on this small magazine, but should put all of my gusto in *Astronomy*. I became *Astronomy* magazine's sixth editor in chief in 2002, and have enjoyed writing about galaxies in one form or another for forty years now.

I still treasure Timothy Ferris's book as one of my early favorites. But practically everything we know about galaxies has changed in the last forty years. In the back of my mind, I always thought about an update of this book for the next generation, now describing the barred spiral nature of the Milky Way, our far deeper knowledge of the galaxies around us in the Local Group, the large-scale structure of clusters and superclusters in the universe, the ubiquitous nature of black holes, and many more topics that we had just hints of forty years ago.

The result is what you hold in your hand. I invite you to join me in our imaginary spaceship and travel far out into the cosmos to explore an amazing universe that we now know in great detail, and one that forty years ago could hardly have been imagined.

THE GHOSTLY GLOW OF THE MILKY WAY
The disk of the Milky Way Galaxy stretches across
our night sky and is visible overhead from a dark
site during the right season and time of night. What
we see is the unresolved light from billions of stars
scattered along our galaxy's disk, as viewed from
within. We can't know exactly what our galaxy looks
like from outside, but only have an approximation
based on mapping the Milky Way's structure.

Chapter One

WHAT ARE GALAXIES?

✳ ✳ ✳

Waves crashed along the beach at Santa Monica, vast stands of forest speckled the mountains north of the city, and a mesmerizing network of roads crisscrossed here and there. In 1923, Los Angeles had a population of one million—just one-quarter of its present size—and the city was in the midst of explosive growth. At the California Institute of Technology, an American physicist, Robert Millikan, won the Nobel Prize in Physics for his measurement of the charge carried by a single proton or electron (the fundamental particles) and for work on the photoelectric effect, including his observation that many metals emit electrons when they are struck by light. Amelia Earhart took periodic flying lessons in the area. The Hollywood Bowl had recently opened for concert

FAST FACT

Galaxies are vast collections of stars, gas, and dust, ensconced in dark matter halos, that are the basic, large-scale building blocks of the cosmos. There are numerous types of galaxies in the universe.

performances. And a young cartoonist named Walt Disney arrived in town with $40 in his pocket.

One big question floated out there: How big is eternity? Is creation limitless?

———

Despite the area's forward-looking involvement in science and technology, it was a primitive time. No one yet knew the size and scope of the universe. People had looked at the brightest galaxies in the sky—the fuzzy patches in Andromeda and the Magellanic Clouds in the Southern Hemisphere—but no one yet understood exactly what they were. One big question floated out there: How big is eternity? Is creation limitless? Soon Los Angeles would play a pivotal role in defining the distance scale of the universe.

THE 100-INCH TELESCOPE

On October 4, 1923, in the midst of this peculiar Western paradise, a brash young astronomer left his Pasadena house and trekked up to the Mount Wilson Observatory, not far from Los Angeles itself, to the 100-inch Hooker Telescope, at the time the largest telescope in the world. Originally from Missouri, Edwin Powell Hubble had moved to Illinois, graduated from the University of Chicago, and then earned a master's degree as a Rhodes Scholar at Oxford University. He embarked on a career in astronomy only after returning to school at the age of twenty-five to pursue a PhD. Hubble was now in his fourth year as a staff astronomer at Mount Wilson. He relished using the 100-inch Hooker Telescope to study his favorite subject: the fuzzy nebulae—mysterious, glowing gas clouds—that appear scattered across the sky.

No one fully understood these nebulae, although they were suspected to be the birthplaces of stars. The adventurous amateur astronomer William Parsons, Third Earl of Rosse, had first sketched nebulae with spiral structures that looked like faintly glowing whirlpool patterns, using his mammoth telescope in rural Ireland in the mid-nineteenth century. But even nearly a century later, little more was known about them. Hubble was interested in cracking the code of nebulae, particularly spiral nebulae. His PhD work had centered on the topic. These nebulae's spiral shapes suggested that they are rotating, but otherwise they mystified Hubble and other astronomers.

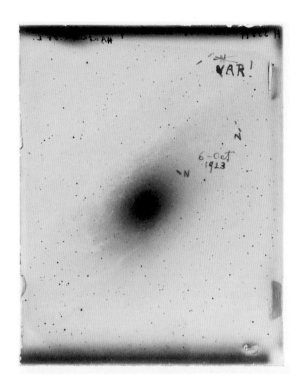

THE PHOTOGRAPHIC PLATE THAT DEFINED GALAXIES

On October 5, 1923, astronomer Edwin Hubble made an exposure of the "Andromeda Nebula" with the 100-inch Hooker Telescope at Mount Wilson Observatory near Los Angeles. He was initially excited, believing he recorded a nova, an exploding star, after analyzing the plate. He marked the star with an "N," and it lies between two tick marks he drew on the glass plate. The famous plate, designated H335H, would help to unlock one of the universe's biggest secrets. A short time later, Hubble realized he had not seen a nova, but a variable star of a particular type called a Cepheid, with very well-known characteristics. Because their absolute magnitudes and light curves were well known, astronomers could use Cepheids to determine distance to them. Astonished, Hubble found that the Andromeda Nebula was actually a distant galaxy, perhaps a million light-years away, far larger than astronomers believed the entire universe was at the time. Also, with this plate, he unlocked the nature of galaxies. (It turns out that the true distance to the Andromeda Galaxy is 2.5 million light-years, we now know.) This plate (left) and an enlargment (below) show the nuclear region of the Andromeda Galaxy on the plate's upper right corner with the famous notation: the "N" for nova crossed out and Hubble's exclamation, in red ink, "Var!"

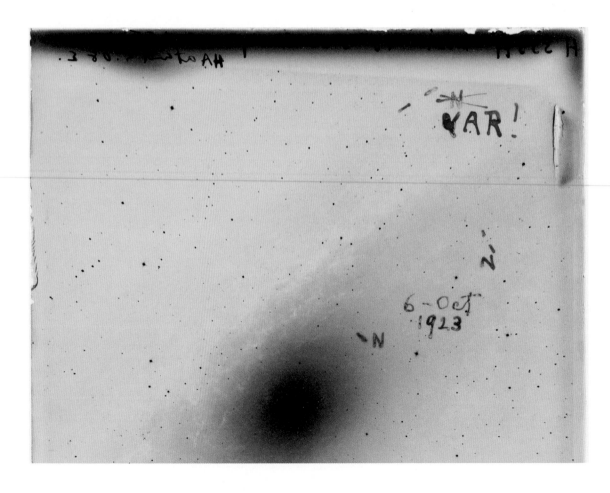

On the night of October 4, 1923, Hubble used the 100-inch Hooker Telescope to take a forty-minute exposure of one of his favorite nebulae, the Great Nebula in the Andromeda constellation. This spiral-shaped cloud was large, bright, and faintly visible to the naked eye as a fuzzy smear of light for those located away from the city lights of Los Angeles. The night had very poor "seeing" when he took the exposure because Earth's atmosphere was relatively turbulent that night, and so the star images were not perfectly small dots. Nevertheless, Hubble's examination of the photographic plate he had made revealed a suspected nova: an exploding star. It was exciting to record such a relatively rare event inside one of the spiral nebulae.

Hubble photographed the Andromeda nebula again the next night, hoping for a better-quality image of the suspected nova. The resulting photographic glass plate, exposed the night of October 5–6 and designated H335H, would become one of the

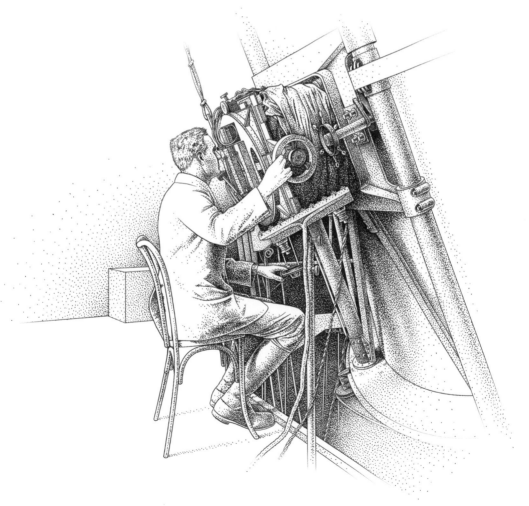

most celebrated in all of scientific history. On it, Hubble successfully recorded the nova again. But before he could analyze it further, his periodic observing run on the 100-inch telescope ended and he had to leave to accommodate other observers. Early in the morning, he left the mountain and returned to Pasadena.

At his office down in Pasadena, away from the mountaintop observatory, he continued studying earlier images of the Andromeda nebula region taken by others. And then he made an unusual discovery. A nova brightens dramatically and then fades into oblivion. But the star he recorded appeared on older plates, brightening and fading regularly over a period of thirty-one days. This star was not a nova, then. It had to be some other kind of star inside the Andromeda nebula.

HUBBLE'S BREAKTHROUGH

Suddenly, Hubble came upon the solution. He realized that he had made an image of a type of star similar to a well-known one in the constellation Cepheus. On his photographic plate H335H, he crossed out "N," for nova, and wrote "VAR!," denoting a variable star. Moreover, this star was a special type of variable that brightened and faded in a precise way. Astronomers had long studied this kind of star, which came to be known as a Cepheid variable (after a star in the constellation Cepheus), and they knew how intrinsically bright it was. By knowing how bright the star really was and measuring how bright it appeared to be in the sky, Hubble could use the star as a guidepost to gauge the distance to it.

> With his photographic plate, Edwin Hubble had single-handedly reset the size of the cosmos.

This was a monumental realization. Hubble calculated that, owing to the star's faint light, it must lie a million light-years away—and so must the entire nebula that surrounds it. This meant that the universe stretches across a distance at least three times larger than most astronomers then believed. With his photographic plate, Edwin Hubble had single-handedly reset the size of the cosmos.

THE 100-INCH
HOOKER TELESCOPE

HOW TO VIEW GALAXIES
from Your Backyard

Soon after Hubble's discovery, astronomers frantically began to collect data on many of the brighter objects in the sky that were suspected of being distant galaxies. These included many of the bright galaxies we can see in our sky today, which can be observed with small telescopes. Here's how to see galaxies in your sky:

✳ **CHOOSE A MOONLESS NIGHT.** You want maximum sky darkness, a clear sky, and an observing site as far away from cities and other light sources as you can get.

✳ **USE A FOUR-INCH OR SIX-INCH TELESCOPE.** This is the minimum size, but an eight-inch or ten-inch telescope is even better, so that you can collect the maximum amount of faint light.

✳ **CHOOSE THE RIGHT TIMING.** The Milky Way is prominent in the summer and winter evening skies, so in spring and autumn other well-known galaxies are unblocked and can be viewed with a backyard telescope. These include the beautiful spiral galaxy M81 and its neighbor M82 in Ursa Major, the Whirlpool Galaxy (M51) in Canes Venatici, M101 in Ursa Major, the Black Eye Galaxy (M64) in Coma Berenices, the Sombrero Galaxy (M104) in Virgo, and the Southern Pinwheel Galaxy (M83) in Hydra.

DISCOVERING THE GALAXIES

Hubble's discovery set off a firestorm of activity among astronomers researching other spiral nebulae. Countless observations followed, and follow-up studies rolled on for many months as bickering and soul-searching lit up the world of professional astronomy. Adding fuel to the fire was a debate staged several years earlier, in 1920, between two prominent astronomers of the day, Harlow Shapley of Princeton University and Heber Curtis of the Allegheny Observatory. Shapley believed that the Milky Way Galaxy constitutes the entire universe, while Curtis speculated that spiral nebulae are separate galaxies from the Milky Way Galaxy—essentially "island universes." Though not everyone would concede it yet, Hubble's discovery seemed to prove that Curtis was right.

Hubble continued imaging Cepheid variables in other spiral nebulae, such as M33 in Triangulum, demonstrating that they, like Andromeda, are so far away that they must be distant galaxies. Hubble's observations indicated that galaxies are the basic units of stars, gas, and dust in the universe, and that they exist on a fantastic scale. He had many doubters, chief among them Shapley, but Hubble pushed on. The findings of the confident thirty-five-year-old were subsequently splashed onto the front page of the *New York Times* by November 1924. Egged on by supporters, he sent a paper summarizing the results to be read at the winter meeting of the American Astronomical Society, the professional organization of astronomers, on New Year's Day 1925. After the distinguished professor Henry Norris Russell of Princeton University read the paper aloud at the gathering, galaxies were on their way to becoming widely accepted.

FAST FACT

The galaxy we live in is called the Milky Way, and it contains some 400 billion stars, one of which is the Sun.

Hubble's observations indicated that galaxies are the basic units of stars, gas, and dust in the universe, and that they exist on a fantastic scale.

A BREAKTHROUGH
WITH GALAXY COLORS

Several more years led to another huge advancement. A galaxy's spectrum is a picture of the collected light from all of its stars and gas. In 1929, Hubble and other astronomers recorded many spectra of galaxies and noticed that they appeared to be shifted toward the red end of the spectrum, increasing the wavelength and lowering the frequency of their light. This was an effect first noticed years earlier, in 1912, by Vesto M. Slipher, an astronomer at the Lowell Observatory in Arizona.

You experience this effect, called a Doppler shift, with sound every time an ambulance with its loud siren passes you. As it approaches, the siren seems high-pitched (because it has a short wavelength and high frequency of sound), and when it passes and heads away from you, the pitch drops lower (increasing the wavelength

GALAXIES SIMILAR TO THE MILKY WAY
Hubble Space Telescope • SDSS

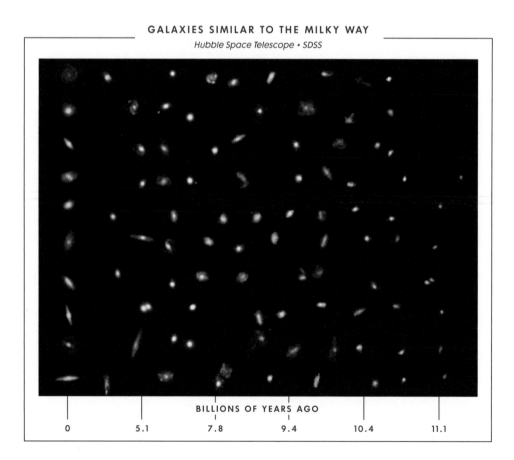

BILLIONS OF YEARS AGO

| 0 | 5.1 | 7.8 | 9.4 | 10.4 | 11.1 |

and lowering the frequency of sound). The same thing happens with light. When objects are moving toward us, the frequency of their light shifts higher, toward the blue end of the spectrum. When they are moving away from us, their light shifts lower, toward the red end of the spectrum. Consequently, the "redshift" of the spectra of distant galaxies indicates that the galaxies are moving away from us. And this means that the universe is not only immensely bigger than previously thought, but also that it is expanding to become even bigger as time marches on.

HERE COMES THE BIG BANG

Hubble's work, building on the earlier studies of Slipher and the astronomer Milton Humason, showed that, generally speaking, all galaxies are moving away from each other over time. Hubble also found that redshifts can be used to calculate distances to galaxies.

This research led to a monumental realization. In 1929, Hubble, with a helpful assist from the Belgian astronomer Georges Lemaître, suggested that the new data he collected about galaxies supported the theory that if traced backward in time, the paths of all of the galaxies led to a small, dense point at which the whole universe began—a "Big Bang" billions of years ago. This Big Bang commenced the expansion that is causing all galaxies to move apart from one another more quickly in space. The whole universe seems to be flying apart.

Hubble analyzed forty-six galaxies and proposed what came to be known as the Hubble Constant, the rate of expansion of the cosmos. He fixed this number as 500 kilometers per second per megaparsec of space, a higher value than what we know is correct today.

Opposite **A SCRAPBOOK PICTURE OF OUR GALAXY'S EARLY YEARS**
With the majestic Hubble Space Telescope, astronomers studied 400 galaxies similar to the Milky Way and created a picture of how our galaxy built up over time. They believe the galaxy began as a low-mass, bluish aggregation of gas, with old stars, that evolved into a flat disk with a central bulge, and later the barred spiral galaxy we know today.

HUBBLE AND THE EXPANDING UNIVERSE

Hubble's credibility skyrocketed following the determination of an expanding universe. This was big stuff: Hubble had piled on lots of supporting evidence for the ideas of the great physicist Albert Einstein, who had proposed in the previous generation that time and space are expanding and that the cosmos is almost unimaginably large.

By the late 1930s, following Hubble's big discoveries, it was becoming clear just how significant galaxies are to the story of the cosmos. Astronomers knew that most of the immensely large universe is filled with darkness. Little matter exists outside the island galaxies, which contain all the bright stuff, the normal matter—stars, gas, dust, and planets. The universe is like a vast and stormy sea, with little ships—galaxies—floating on it and a virtually limitless ocean of utter darkness and foreboding void in between them.

CLASSIFYING GALAXIES

By this time, Hubble understood the broad types of galaxies, which he classified in a "tuning fork" diagram. There are "spiral galaxies," like Andromeda, and "barred spirals," which are similar to spiral galaxies but contain a rectangular "bar" of material through their centers. There are "ellipticals," spherical masses of stars, gas, and dust. There are "lenticulars," which appear lens-shaped, and there are "irregulars," relatively formless clouds of matter lacking organized structure. In the late 1930s, astronomers discovered examples of a new class, "dwarf spheroidal galaxies," and later still astronomers found so-called "peculiar galaxies," which seem highly distorted. By the end of the 1950s, they had devised an improved way to classify galaxies, based on the research of the French astronomer Gérard de Vaucouleurs of the University of Texas.

FAST FACT

Some galaxies are classified using their characteristics, or behaviors. For example, interacting or merging galaxies are pairs or groups of galaxies that are intertwined in a cosmic dance.

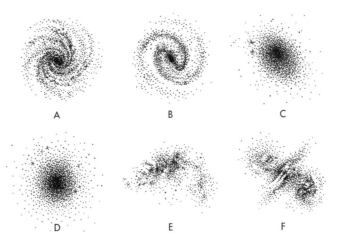

Examples of all these types of galaxies can be viewed with a telescope from a dark sky site. They include:

✳ **Spiral galaxies (A):** the Sunflower Galaxy (M63), IC 342, and NGC 1232

✳ **Barred spiral galaxies (B):** NGC 1300, NGC 1512, NGC 1530, NGC 4921, and NGC 5701

✳ **Elliptical galaxies (C):** M49, M87, and NGC 1052

✳ **Lenticular (lens-shaped) galaxies (D):** M84, NGC 2787, and NGC 4111

✳ **Irregular galaxies (E):** NGC 1569, NGC 3239, and NGC 4214

✳ **Peculiar galaxies (F):** Arp 81, Arp 220, Centaurus A, Fornax A, M82, and Perseus A

De Vaucouleurs's classification scheme was more complex, forming a 3-D "cosmic lemon" that accounted for more properties of the basic types of galaxies. For spiral galaxies, this included further details on bars, notes on whether a galaxy showed rings of encircling material, and notes on how tightly or loosely wound the arms of a spiral galaxy appeared to be. De Vaucouleurs also cataloged details about irregular galaxies and described peculiar galaxies, which had experienced galactic train wrecks—interactions with nearby galaxies that warped their shapes.

How Astronomers

CLASSIFY GALAXIES

Astronomers have struggled to classify galaxies since the 1920s, when Edwin Hubble at California's Mount Wilson Observatory first proved that they were "island universes" beyond our Milky Way. Hubble found that he could lump most galaxies into a few broad types—elliptical, normal spiral, and barred spiral—and he placed them into a tuning fork-shaped sequence.

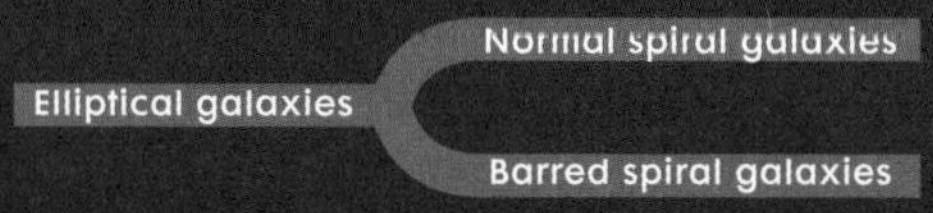

Galaxies display a huge range of sizes and masses—from nearby dwarfs that contain fewer stars than a globular cluster to mammoths like M87. Galaxies show a range of structures, too, including bulges, disks, bars, rings, and spiral arms.

Elliptical galaxies

Normal spiral galaxies

Barred spiral galaxies

The subtler details of galactic structure required a more nuanced scheme than Hubble's. In the late 1950s, Gérard de Vaucouleurs devised a 3-D version of the tuning fork—shown opposite—based on observations of a few hundred southern galaxies.

But distant galaxies, seen when the universe was younger, smaller, and more crowded, don't fit into the scheme. Moreover, new surveys of tens of thousands of galaxies suggest they occupy two distinct zones when plotted by color and brightness. No one yet knows how to reconcile such observations with the smooth variations astronomers see among different galaxy types.

Study the accompanying diagram carefully. Giving Hubble's tuning fork a half twist creates the lemon-shaped "classification volume" Gérard de Vaucouleurs developed beginning in the late 1950s. Normal galaxies occupy the top half, barred galaxies go on the bottom. Different galaxy types occur left to right. Codes describe each galaxy based on its position within the volume. The "lemon's" height reflects the relative number of each galaxy type found in the sample.

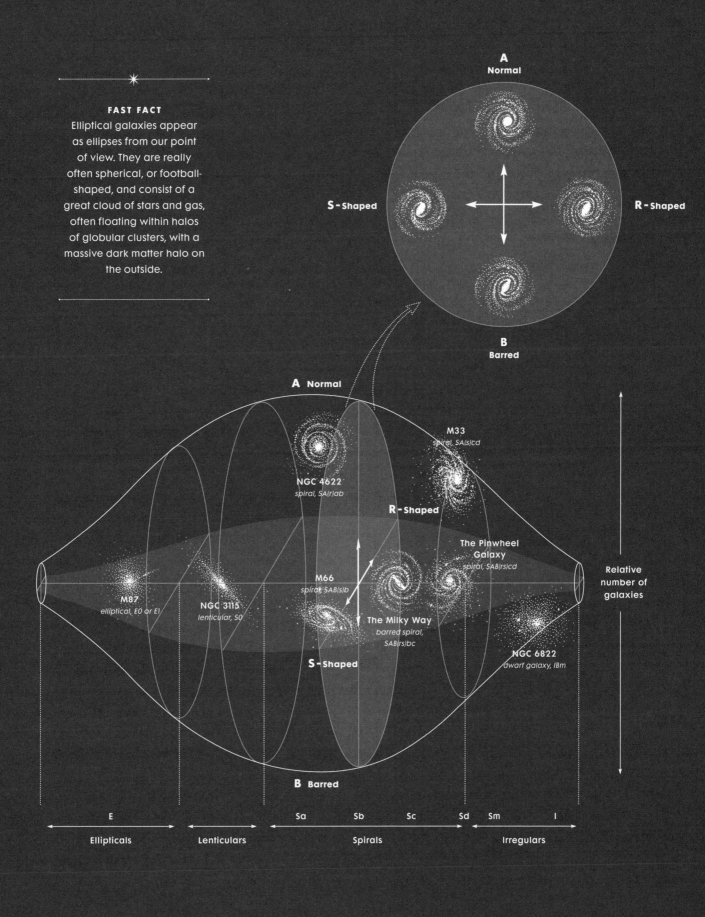

A Normal

S-Shaped

R-Shaped

B
Barred

A Normal

NGC 4622
spiral, SA(r)ab

M33
spiral, SA(s)cd

R-Shaped

The Pinwheel
Galaxy
spiral, SAB(rs)cd

M66
spiral, SAB(s)b

The Milky Way
*barred spiral,
SAB(rs)bc*

M87
elliptical, E0 or E1

NGC 3115
lenticular, S0

S-Shaped

NGC 6822
dwarf galaxy, IBm

B Barred

Relative
number of
galaxies

E	Lenticulars	Sa	Sb	Sc	Sd	Sm	I

Ellipticals   Lenticulars   Spirals   Irregulars

THE INCREDIBLE
SCOPE OF THE COSMOS

For years, astronomers have quoted the results of deep galaxy surveys that suggest something like 100 billion galaxies exist in the universe. A 2016 study suggests that the total number of galaxies could be *2 trillion*. But that study looks back into the early universe, and many galaxies have merged over time, creating a smaller "current" number like 100 billion. We are hanging out in just one of them, the Milky Way.

These basic structures of the cosmos, like ships floating on an ocean of vast darkness, give us a glimpse beyond our world to understand the meaning of why we're here.

As astronomers have discovered more and more galaxies since the 1920s, they have acquired one really fundamental piece of knowledge. The universe is *really big*! Let's imagine that you could climb into a spaceship and travel out into the universe, seeing more and more distant things as you went on. Let's also imagine that the spaceship could travel at the speed of light—the fastest speed we know of in the cosmos. That's about 186,000 miles per second, the speed at which photons, particles of light, are striking your eyes, enabling you to read this book. (Photons can travel that fast because they have no mass, and spaceships have mass—so we know that spaceships couldn't move that fast. But for the sake of understanding the size of the universe, let's pretend that our spaceship could.)

FAST FACT

The most familiar type of galaxy, a spiral galaxy, has spiral arms that wind around the galaxy's center in a glowing disk. The spiral arms are created by density waves—ridges that are denser than the areas surrounding them. They contain a disk of stars and gas, a central hub or bulge packed with stars and gas, a large halo of globular star clusters, and an outer halo of dark matter.

Galaxies, the basic structures of the cosmos, are like ships floating on an ocean of vast darkness.

TRAVELING
TO THE GALAXIES

In our spaceship, let's set out from the Milky Way Galaxy, our home. The closest galaxy we can encounter is the Sagittarius Dwarf Spheroidal Galaxy, a tiny galaxy that orbits ours. If we move at the speed of light, it would take us 70,000 years to reach this galaxy. Another way of thinking about these enormous distances is to understand how long the light we see from other galaxies now has been traveling through space to reach us. The light from the Sagittarius Dwarf Spheroidal Galaxy has traveled since humans made their earliest bits of art inside caves in South Africa. If we traveled for 163,000 years in our spaceship, we could arrive at the Large Magellanic Cloud, our galaxy's largest satellite. Traveling for 200,000 years would carry us to the Small Magellanic Cloud, another satellite of our Milky Way. The light you see from this galaxy tonight has traveled through space since our earliest human ancestors closely linked to our species walked the African plains.

FAST FACT

Clusters of galaxies are collections of galaxies held together by common gravity. They contain a dozen or more large spirals or ellipticals, and possibly hundreds of smaller galaxies. Galaxy clusters typically span several tens of millions of light-years.

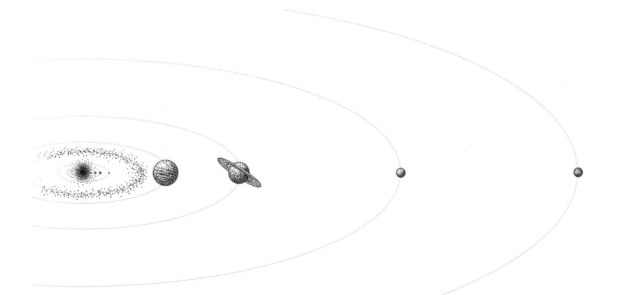

Vast to us, the distances in the solar system pale compare to journeys to other galaxies.

But those are dwarf galaxies that are very close to us. The largest nearby galaxy is the Andromeda Galaxy, which would take us 2.5 million years to reach in our spaceship. The light you see from this galaxy tonight has traveled through space since some of our earliest human ancestors were here on Earth.

And these are just some of the closest galaxies to us. Traveling outward, you would find countless examples of strange and beautiful galaxies at all manner of distances. These would include beautiful spirals like IC 239, M100, M106, NGC 210, NGC 2683, NGC 2841, NGC 3310, NGC 3338, NGC 4565, and NGC 6946. You would encounter fields of multiple galaxies like those in the Leo Trio (M65, M66, and NGC 3628), M81 and M82, or the galaxy group Hickson 31. Some galaxies that seem to be connected, like NGC 3314, would grow away from each other as you approached and their visual alignment disappeared. You would encounter numerous weird, distorted galaxies, the result of interactions or disruptions by black holes, like Arp 188, ESO 243-49, NGC 474, NGC 660, NGC 2685, NGC 4622, NGC 5291, NGC 7714, and UGC 697.

As you read this book, you'll see how enormous the cosmos is and understand that, fundamentally, it is filled with galaxies. You'll encounter Virgo Cluster galaxies that would take 50 million years to reach in our light-speed spaceship. More distant galaxies are arranged in clusters and superclusters that we can see from Earth, and some lie hundreds of millions or billions of light-years away. So reaching the most distant galaxies we can see would take us more than 13 billion years, traveling at the fastest speed we know of—a speed at which only particles of light can go. Living our lives on this third planet from the Sun in our solar system, it's easy to ignore how unbelievably immense the universe is. But moving farther and farther out into the universe to explore galaxies allows us to understand how the universe came to be, and where it's going.

FAST FACT
Even larger than clusters of galaxies are galaxy superclusters. These enormous aggregations can hold 100 galaxy groups, or clusters, and stretch over half a billion light-years across.

Opposite **NGC 7424: A SPLENDID FACE-ON BARRED SPIRAL GALAXY**
If we could see the Milky Way from a great distance, face-on to our line of sight, it would look very much like this galaxy, NGC 7424 in the southern constellation Grus. Lying 40 million light-years away, NGC 7424 stretches 100,000 light-years across, the same size as the Milky Way's disk. Numerous clusters of massive stars and pinkish HII regions, areas of new star birth, lie scattered along its arms.

Previous **THE ANDROMEDA GALAXY IN**
ULTRAVIOLET LIGHT
Imaged in high-energy ultraviolet
wavelengths, the Andromeda Galaxy
appears to have spiral arms that look like
rings. This is due to highly energized light
from young, massive stars peppered across
the arms. The intense star formation is one
line of evidence of an encounter with the
satellite galaxy M32, visible as a fuzzy knot
just above the spiral arm, above and left of
M31's nucleus.

Above **A CENTRAL SLICE OF THE**
ANDROMEDA GALAXY
This amazing Hubble Space Telescope
mosaic of the Andromeda Galaxy resolves
individual stars in the object, despite its
great distance of 2.5 million light-years. The
innermost hub of the galaxy appears at left,
and a slice of spiral arm on the right side.
Areas of young, blue stars show regions of
recent star formation.

Overleaf **THE ANDROMEDA GALAXY IN**
BLACK AND WHITE
A monochromatic image of the Andromeda
Galaxy, our famous galactic neighbor, reveals
the intricate details of its spiral arms, areas of
swirling gas clouds near the galaxy's center,
and two satellite companions, M32 (above left
of the center of Andromeda), and NGC 205
(below the galaxy's center).

NGC 266: A BARRED SPIRAL GALAXY WITH AN ENERGIZED NUCLEUS

The unusual barred spiral NGC 266 lies in the constellation Pisces, at a distance of some 215 million light-years away. It is a LINER type galaxy, meaning it has a bright, active nucleus that is casting out energy from a central black hole.

Above **NGC 6744: A GALAXY THAT LOOKS LIKE THE MILKY WAY**
The brilliant galaxy NGC 6744, lying in the southern sky in Pavo, is a larger version of the Milky Way. This barred spiral stretches 175,000 light-years across; some 75 percent bigger than our home. Its structure, however, is similar to ours, with a core, a strong bar through its center, and radiating spiral arms filled with glowing stars and gas.

Overleaf **THE MAGNIFICENT EDGE-ON SOMBRERO GALAXY**
One of the greatest edge-on galaxies in the sky, and the one most people say looks like a flying saucer, is the Sombrero Galaxy (M104) in Virgo. This galaxy consists of a great rotating disk with a prominent dust lane edging it, consumed by a glowing halo of gas and stars. It lies 29 million light-years away and is about half the size of the Milky Way, at 49,000 light-years.

43

Opposite **NGC 1569: A NEARBY GALAXY OFFERS A STARBURST BLAST**
The dwarf irregular galaxy NGC 1569 lies in the constellation Camelopardalis at a distance of 11 million light-years. This small galaxy is undergoing a large, sustained starburst event: during the last 100 million years, it has formed stars at a rate 100 times greater than the Milky Way. Numerous, brilliant blue star clusters in the galaxy are young and hot, and many supernovae have flared in this galaxy, creating distinctive bubbles of gas.

Above **NGC 2787: CLOSE-UP PEEK AT A LENTICULAR GALAXY**
The lenticular galaxy NGC 2787 in Ursa Major is one of the better-known examples of this lens-shaped class of galaxies. It lies 24 million light-years away and shows an intensely bright core surrounded by a halo of stars and gas, around which are wrapped tightly wound lanes of dust. This galaxy has a central black hole with about the same mass as the one in the Milky Way.

Overleaf **NGC 1300: A STRIKING FACE-ON BARRED SPIRAL**
The barred spiral NGC 1300 in Eridanus is a "grand design" galaxy with well-defined arms and a prominent, bulgy bar. It is about the same size as the Milky Way, and lies some 60 million light-years away. The supermassive black hole in the galaxy's center has nearly twice the mass of the Milky Way's, weighing in at 7.3 million solar masses.

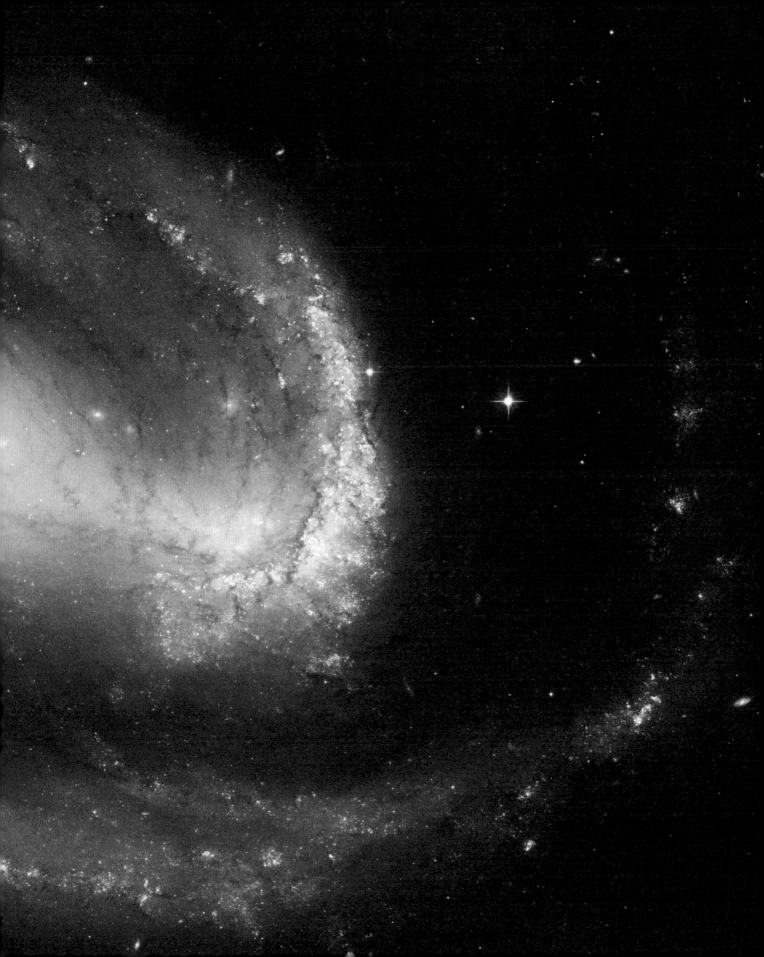

**NGC 1530: A BARRED SPIRAL GALAXY WITH A
"MINI-SPIRAL" NUCLEUS**
Oriented nearly face-on to our line of sight, the barred
spiral NGC 1530 in Camelopardalis lies some 80 million
light-years away. Its very prominent bar connects with
large, well-defined spiral arms. The galaxy's center has
a swirling pattern reminiscent of the spiral form of a
galaxy in itself.

**NGC 3239: A WARPED, IRREGULAR GALAXY
SHOWING ITS SUPERNOVA**

The strange object NGC 3239 in Leo is an irregular galaxy
showing a weird, twisting pair of arm-like extensions,
suggesting a major encounter with another galaxy in the
past. At a distance of 25 million light-years, this galaxy
spans 40,000 light-years across. The bright star just above
the galaxy's center is a foreground star, but just below
it and to its right lies Supernova 2012A, which for a brief
time winked on as this aged star died.

Opposite **THE WEIRD BARRED SPIRAL NGC 4921**

The barred spiral galaxy NGC 4921 in Coma Berenices is a distant object, at 320 million light-years. It is a so-called anemic galaxy, labeled as such by Canadian astronomer Sidney van den Bergh, because it has a very low rate of star formation. Its lovely spiral pattern around the small central bar gives it the appearance of a delicate painting.

Top **THE ELEGANT FACE-ON BARRED SPIRAL GALAXY NGC 5701**

NGC 5701 gives us a view of a barred spiral that is fairly similar to the Milky Way, lying at a distance of 77 million light-years in the constellation Virgo.

Bottom **INFANT STAR CLUSTERS RING THE GALAXY NGC 1512**

Barred spiral galaxy NGC 1512 lies some 38 million light-years away in the southern constellation Horologium. Its intensely yellow colored disk is ringed by a group of bluish infant star clusters, and contains a bar that is too faint to be seen. Astronomers believe the bar funnels gas into the outer ring, leading to the bonanza of star formation.

**THE SUNFLOWER GALAXY'S TIGHTLY
WOUND SPIRAL ARMS**
Sometimes called the Sunflower Galaxy,
M63 in Canes Venatici lies at a distance
of 27 million light-years. It is bright and is
a favorite target for backyard observers.
M63 is a flocculent galaxy, with patchy,
indistinct arms. It is also a so-called LINER
galaxy, with an active nucleus powered by
a supermassive black hole.

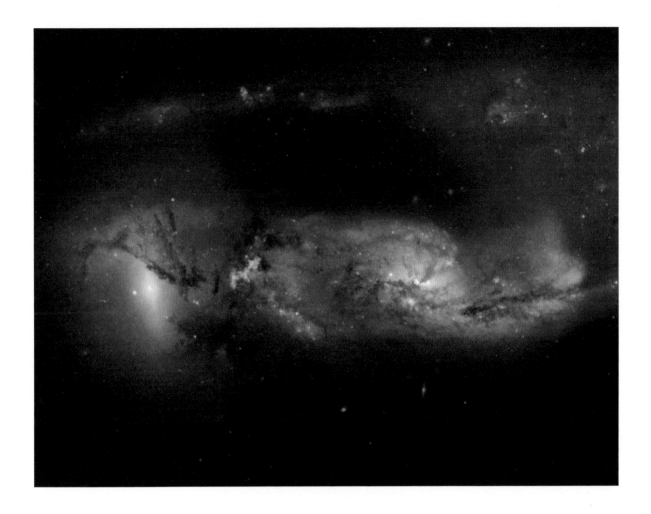

Opposite NGC 1073, A DELICATE
FACE-ON BARRED SPIRAL GALAXY
This lovely barred spiral floats at a
distance of 55 million light-years and
is located in the constellation Cetus. It
features a large central bar and,
unlike the Milky Way, has relatively
asymmetrical spiral arms.

Above ARP 81: WORLDS COLLIDE IN A
GALACTIC HEAD-ON CRASH
Two distorted galaxies in Draco are
locked in a tidal embrace: They are NGC
6622 (left) and NGC 6621, collectively
known as Arp 81. This composite reveals
twisted streams of gas and stars, chaotic
star formation, and a massive tidal tail
stretching across the top portion of the
picture. The tail measures 200,000 light-
years across, twice the diameter of the
Milky Way. These galaxies lie at a distance
of 280 million light-years.

Top **THE UNUSUAL BARRED SPIRAL GALAXY IC 239**
IC 239 lies at a distance of 46 million light-years in the constellation Andromeda. Flanked by bright foreground stars, the galaxy makes an entrancing sight in backyard telescopes.

Bottom **BARRED SPIRAL GALAXY NGC 210 AND ITS LENS-SHAPED CENTER**
NGC 210, a bright barred spiral galaxy lying 67 million light-years away in Cetus, shows a very bright core that is lenticular in shape. The bar is relatively difficult to spot, and the galaxy's hazy spiral arms suggest it may be evolving toward becoming a ring galaxy.

Opposite **NGC 1398: ANOTHER MILKY WAY ANALOG**
The beautiful barred spiral NGC 1398 lies in the southern constellation Fornax, and provides another analog to the Milky Way's structure. This galaxy is somewhat larger than ours, with a diameter of 135,000 light-years, and lies some 65 million light-years from Earth.

Overleaf **THE SKY'S BEST EDGE-ON GALAXY, A SLENDER NEEDLE OF LIGHT**
NGC 4565 in Coma Berenices is the brightest and most prominent galaxy in our sky that is oriented perfectly edge-on to our line of sight. We see its disk as a thin, silvery needle. Some 43 million light-years off, it lies in the Virgo Cluster and has a prominent central bulge, suggesting it may be a barred spiral.

THE LEO TRIO: M65, M66, AND NGC 3628
A low-power eyepiece field of view trained on
the right portion of the constellation Leo nets
three galaxies for the price of one, the Leo Trio,
or Leo Triplet. M65 (top right) is a barred spiral
some 35 million light-years away. M66 (bottom
right), also a barred spiral, lies 36 million light-
years distant. NGC 3628, an edge-on galaxy
with a prominent dust lane (lower left), is a
spiral that lies 35 million light-years away. The
three are the brightest members of a small
group of galaxies.

NGC 3338: A GALAXY WITH MAJESTIC, SWIRLING SPIRAL ARMS
Located at a distance of 80 million light-years in the constellation Leo, NGC 3338 shows bright, highly inclined spiral arms wrapping away from an intense, oval-shaped nucleus.

Opposite **A COSMIC BLENDER OF MULTIPLE SHELLS AND TAILS**
The weirdly distorted elliptical galaxy NGC 474 in Pisces lies at a distance of 100 million light-years. The neighboring spiral galaxy NGC 470 lies just above it. Multiple shells and tidal tails surround NGC 474, caused by interactions with its neighbors and by density waves that propagate through the medium. This mammoth object stretches 250,000 light-years across—2½ times the diameter of the Milky Way.

Above **NGC 2683: A STRIKING EDGE-ON SPIRAL GALAXY**
A slender needle of light, spiral galaxy NGC 2683 in Lynx is oriented nearly edge-on to our line of sight, and makes a wonderful telescopic target. It may be a barred spiral; galaxies aligned like this are hard to classify, as we can't see their disks well. The galaxy has a remarkably bright core due to a large population of old, yellowish stars, and it lies some 20 million light-years away.

Above NGC 5291: AN ANCIENT GALAXY COLLISION CREATES A SEASHELL

The bright yellowish galaxy in the center of this field in Centaurus is NGC 5291, sometimes called the Seashell Galaxy. It is the result of a galaxy merger and is highly distorted; the tidally disrupted galaxy below it is being shredded and will merge fully with the larger object. This pair of galaxies lies in the rich galaxy cluster Abell 3574, other members of which lie in the field. The interacting pair lies 200 million light-years away.

Opposite A FIREWORKS GALAXY: FACE-ON SPIRAL NGC 6946

The speckled face of NGC 6946 in Cygnus, right on the border with Cepheus, suggests a fireworks display of color. Bluish spiral arms surround a yellowish central hub, and are speckled with bright pinkish regions of active star formation. The galaxy lies at a distance of 22 million light-years and has frequently served up supernovae, exploding stars. Ten bright supernovae have appeared in this galaxy between 1917 and 2017.

Opposite THE HIGHLY DISRUPTED GALAXY NGC 7714

NGC 7714 is a spiral galaxy in Pisces that has been smashed by an encounter with its neighbor NGC 7715 (not visible). The latter galaxy presumably shot through this object like a missile, leaving a warped disk and giant ring of stars lying in its wake. The galaxies lie at a distance of about 100 million light-years.

Above THE HIGHLY DISTORTED GALAXY ARP 188 AND ITS TADPOLE TAIL

Sometimes called the Tadpole Galaxy, Arp 188 is an interacting galaxy in the constellation Draco. It lies at a distance of 400 million light-years and shows a long streamer that indicates gravitational interaction with one or more galaxies in the distant past. The tail stretches more than 280,000 light-years long and features bright blue knots of star formation.

69

Above NGC 4622: MARCHING TO THE BEAT OF A DIFFERENT DRUMMER

An unusual galaxy in Centaurus, NGC 4622, is sometimes called the Backward Galaxy. It is a rare example of a galaxy with so-called leading arms: in most galaxies, the spiral arms trail the motion of the disk. In NGC 4622, however, the spiral arms lead the rotation of the galaxy. This may be due to gravitational interaction between the galaxy and another, smaller object. NGC 4622 lies at a distance of 110 million light-years.

Opposite NGC 3314: A CHANCE ALIGNMENT OF GALAXIES ON THE SKY

The universe is a big place, and the sky filled with darkness, but sometimes things align just right. NGC 3314 in Hydra appears to be an intertwined pair of galaxies, but it is in fact merely a chance alignment of two objects at different distances. A face-on foreground galaxy, NGC 3314a, lies at a distance of 117 million light-years, directly superimposed on the center of the background galaxy, NGC 3314b, which lies 140 million light-years distant.

Overleaf NGC 2841: A FLOCCULENT SPIRAL GALAXY

The galaxy NGC 2841 in Ursa Major shows arms that are patchy and discontinuous, making it a so-called flocculent galaxy. This high-resolution image of the galaxy's central region was made with the Hubble Space Telescope, which determined the galaxy lies 46 million light-years away.

Chapter Two

INSIDE THE MILKY WAY GALAXY

✳ ✳ ✳

G o out on a warm summer evening, away from city lights, and after your eyes adapt to the darkness, you'll see the faintly glowing band of the Milky Way stretching from Cygnus overhead down to Sagittarius far below.

The band you see is the Milky Way, which is a cluster of light from billions of unresolved stars. We perceive it as a bright band because we're inside it. In reality, the Milky Way is a barred spiral galaxy, and our Sun just one of the several hundred billion stars it contains.

Our Milky Way Galaxy consists of a disk of stars, including the Sun; a radiant center containing a giant black hole and swirling stars and gas; and outer reaches far beyond us containing old stars in scattered clusters and a gigantic halo of dark matter. Moreover, the Milky Way has a number of satellite galaxies that orbit us, held in place by gravity. We'll explore all of the components of the Milky Way in the following pages.

FAST FACT
Traveling at 186,000 miles per second, the fastest speed there is, light takes at least 100,000 years to travel from one end of the Milky Way to another.

THE SLOW REALIZATION OF GALAXIES

In the fourth century BC, the Greek philosopher Aristotle recorded his thought that the Milky Way might consist of distant stars. But it was one of the fathers of observational science, Galileo, who first turned his newly created telescope toward the glow of the Milky Way, in the autumn of AD 1609. He was the first to see that this nebulous band of light was made up of countless stars. Later on, in 1755, the German philosopher Immanuel Kant speculated that the Milky Way could be a huge rotating body of stars held together by gravity. Kant also introduced the term "island universes" to describe large bodies of stars. This term was used extensively until Hubble's discovery.

FAST FACT

The first person to see the nature of the Milky Way, the luminous band of light in our sky, was Galileo, who turned his new telescope toward it in AD 1609. He saw that it consists of innumerable stars.

The nature of the Milky Way came into focus pretty rapidly with Hubble's discovery in 1923 and the subsequent resolution of the debate over the size of the universe between the astronomers Harlow Shapley and Heber Curtis. By the end of the 1920s, nearly everyone agreed that the Milky Way is our home galaxy and that the universe contains a vast number of other galaxies stretching out over immense distances.

But of course, it's really hard to figure out the nature of a galaxy when you're inside it. So throughout most of the twentieth century, the tasks undertaken by astronomers to understand the Milky Way included mapping, counting stars, and measuring distances in order to piece together a vision of the galaxy's form and structure.

Several years later, a different group of astronomers, using the Spitzer Space Telescope, an orbiting observatory that watches the cosmos in infrared light, added to the clarity of our galaxy's structure. One of the biggest roadblocks to understanding our galaxy is dust: the galaxy contains vast amounts of dust that block the light from stars and other objects beyond. Infrared light, however, allows us to see through this dust and study the structures that lie very far away. With an instrument attached to the Spitzer Space Telescope called GLIMPSE—short for Galactic Legacy Mid-Plane Survey Extraordinaire—astronomers have been able to map our galaxy in the most accurate way to date. By counting stars in the imagery produced by this instrument, the GLIMPSE team confirmed the existence of the Milky Way's central bar when they published their results in 2005. This was a big step: the discovery that we do not live in a simple spiral galaxy, like Andromeda, but in a barred spiral. Thanks to the GLIMPSE project and other recent studies, we now have a reasonably accurate picture of the structure and the contents of the Milky Way Galaxy for the first time in history.

THE PARTS OF THE MILKY WAY

A Barred Spiral Galaxy

As with all galaxies, the Milky Way has a large number of components. Edwin Hubble showed us that galaxies are islands of stars, gas, and dust, but of course these islands vary immensely from galaxy to galaxy. Our galaxy is a moderately large barred spiral. Though it's one of three large galaxies in the galaxy group to which we belong, it's dwarfed by the largest galaxies in the universe. Spiral galaxies have a prominent disk of stars and gas that rotates around like a compact disc, spinning again and again. Barred spirals have this disk but are notable for the prominent bar through their centers. The spiral arms originate from this bar rather than from the galaxy's center.

The Galactic Core

The Milky Way's core is the downtown area, the most congested and active region of the galaxy and the center of it all. As we see it in our sky, the direction of the galactic center lies in the constellation Sagittarius, the exact spot being a short distance west of the midpoint of a line between two famous deep-sky objects: the bright open cluster M7 and the Lagoon Nebula (M8).

The core of our galaxy contains a dense region of stars and gas, swirling around a very massive central object. First detected as a radio source and designated Sagittarius A* (pronounced "Sagittarius A-star"), this object was discovered to be a supermassive black hole. By carefully measuring the velocities of the stars flinging around the galaxy's center, researchers estimated that the mass of the

FAST FACT
The Milky Way is approximately 9 billion years old and probably consists of the merged components of dozens of earlier protogalaxies—perhaps as many as 100. The oldest stars in the galaxy, however, are more than 13 billion years old.

Opposite Scientists think our galaxy has four major spiral arms that wind their way out from a central bar. The Sun lies approximately 26,000 light-years from the center in a minor arm, the Orion Arm.

THE MILKY WAY GALAXY

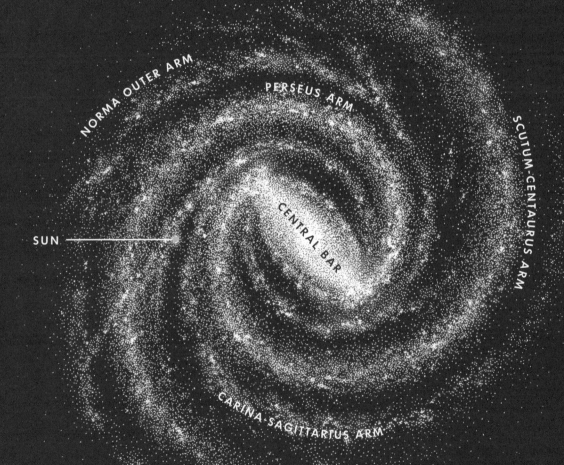

NORMA OUTER ARM

PERSEUS ARM

SCUTUM-CENTAURUS ARM

CENTRAL BAR

SUN

CARINA-SAGITTARIUS ARM

black hole is 4.3 million times the mass of the Sun. And all of that mass is squeezed into a sphere about the size of Mercury's orbit around the Sun. It makes for an incredibly weird and densely packed environment, with stars and gas clouds flying around it at great speeds.

The Galactic Bulge

Surrounding the Milky Way's core is what astronomers term the central bulge. This area also has a very high density of stars, gas, and dust. The bulge contains some 20 billion times the mass of our Sun and shines with 5 billion times the luminosity of the Sun.

The Milky Way's Disk

Most of the stars and gas in the Milky Way lie within the galaxy's bright disk, a glowing component that is roughly analogous in shape and motion to a slowly spinning CD. The stellar disk extends out about 44,000 light-years on either side of the galaxy's center; beyond this distance, the disk continues, but the population of stars drops off significantly. Overall, the diameter of the Milky Way's bright disk is at least 100,000 light-years, though forthcoming research suggests that its diameter could be significantly larger than what was previously believed.

The galaxy's disk of stars, gas, and dust has two components, a thin disk and a thick disk. Astronomers know that the thin disk contains some 90 percent of the stars in the galaxy, including the Sun and our solar system and all of the young, massive stars that have been born in open star clusters. The thin disk is younger, having formed more than 8 billion years ago. The thick disk is a little older and contains older stars.

This thin disk that contains most of the galaxy's stars is some 1,500 light-years thick. It acts like a slowly spinning platter, going round and round and round. Encapsulating this disk is the so-called thick disk, some 3,000 light-years in depth, which contains a much smaller number of stars. Stars in the thick disk are older, having formed earlier in the galaxy's history, and most of the gas and dust in the galaxy exists in a thin layer that is no more than 500 light-years from the disk.

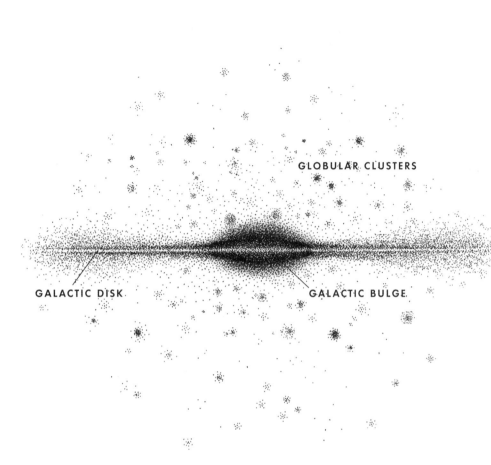

GLOBULAR CLUSTERS

GALACTIC DISK

GALACTIC BULGE

At least 158 dense balls of mostly
ancient stars called globular clusters
orbit the center of the Milky Way
within its extended halo.

The Milky Way's Bar

Within the galaxy's disk, the Milky Way's central bar consists of two elements. What astronomers call the Central ("Bulgy") Bar extends about 11,400 light-years from the galaxy's center. Extending over a larger diameter and distance is the Long Bar, which surrounds the Central Bar and has a diameter of some 28,700 light-years.

FAST FACT

The galaxy's disk is about
1,000 light-years thick, top to
bottom. The galaxy's central
bar stretches some 10,000
light-years across.

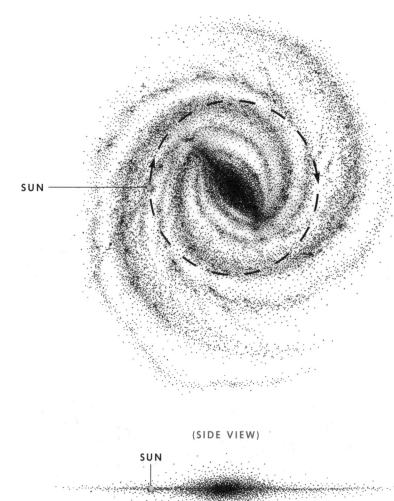

SUN

(SIDE VIEW)

SUN

Spiral Arms

The Milky Way has at least six main spiral arms, which unwind outward from the galaxy's center. Let's examine them, moving from the innermost to the galactic center to the outermost, to see what makes them different.

The innermost spiral arm is the 3-kpc, which was discovered in the 1950s by astronomers making radio observations. It contains about 10 million solar masses of gas, mostly in the form of hydrogen atoms and molecules. As it unwinds, its outer

Opposite Our Sun and solar system rotate around the Milky Way's center at 450,000 miles per hour, completing one orbit in about 220 million years.

parts are termed the Perseus Arm, which is one of the two most prominent arms of the galaxy.

Next come the Norma and Outer Arms. The Norma Arm lies near the galaxy's center, and the outer portion of this structure is the Outer Arm.

Moving outward, we encounter the Scutum-Centaurus Arm—a long, diffuse streamer of stars and gas that originates from one end of the galaxy's bar. This area is rich in star-forming regions.

Next comes the Carina-Sagittarius Arm. Although it is a relatively minor arm, it is rather easy to trace, as it contains a number of star-forming regions that light up portions of it along its form.

Two additional arms, or spurs, are somewhat less pronounced in structure, but are important to us because of where they lie. In between the Carina-Sagittarius and Perseus Arms lies a short arm, or spur, called the Orion-Cygnus Arm. This short arm contains the Sun and our solar system, and so could be called the most important arm of the galaxy to us. It's called the Orion-Cygnus Arm because some of the brightest stars in these familiar constellations lie within it.

The Milky Way's Halo

Far out from the galaxy's disk, both thin and thick, is what astronomers call the "halo," which wraps it all up. This outer element contains metal-poor "globular star clusters"—globe-shaped spheres of old, yellowish stars—and you can see many of them quite easily through a backyard telescope. Among the brightest are the Hercules Cluster (M13) and Omega Centauri. The halo also contains clouds of so-called neutral hydrogen gas—made up of hydrogen atoms with one proton and one electron—and large amounts of dark matter, and it reaches to some 200,000 light-years from the galactic center on all sides.

The Orion-Cygnus Arm contains the Sun and our solar system.

THE MILKY WAY'S SATELLITE GALAXIES

A number of smaller galaxies in our vicinity in the Milky Way orbit us as satellites. The two most famous were discovered a thousand years ago but are named after the Portuguese explorer Ferdinand Magellan because they were written about in the ship's logs for his voyage in 1519. The Magellanic Clouds are visible in the Southern Hemisphere sky. They look like bright, detached patches of the Milky Way to the unaided eye but in fact are separate galaxies orbiting ours, and each has an unusual makeup.

The Large Magellanic Cloud (LMC), lying some 163,000 light-years away in the constellations Dorado and Mensa, is an irregular galaxy with a slightly barred structure. At about 14,000 light-years across, it's some 14 percent as large as the Milky Way, and it contains about 10 billion solar masses of material. Its form is disrupted by tidal forces exerted on it from the Milky Way. The Large Magellanic Cloud is rich in gas and dust and contains the Tarantula Nebula, an incredibly active star-forming region that is easily visible through backyard telescopes. It was also the site of an exploding star, Supernova 1987A, some thirty years ago, the nearest such exploding star that astronomers have seen in many years.

Second in size to its sister galaxy, and also in the Southern Hemisphere sky, is the Small Magellanic Cloud (SMC), located in the constellations Tucana and Hydrus. Also an irregular galaxy showing features of a barred spiral, this galaxy is fainter than the LMC and lies slightly farther away, at 200,000 light-years. It's a mere 7,000 light-years across and has about 7 billion solar masses of material. A faint bridge of gas called the Magellanic Stream connects these two smaller galaxies and shows evidence of the gravitational tug of the Milky Way on both of them.

FAST FACTS

Our own group of galaxies, the Local Group, contains at least 55 galaxies, including the Milky Way, and perhaps as many as 100 galaxies, including many faint dwarfs.

The Sagittarius Dwarf Elliptical Galaxy, a satellite of the Milky Way, is a small galaxy that is being torn apart by the immense gravity of our galaxy. It will eventually be absorbed into the Milky Way.

Opposite **THE LOCAL BUBBLE**
The Local Bubble is a void in the interstellar medium where the density of hydrogen atoms is only one-tenth that of the rest of the Milky Way. What gas remains is hot and emits X-rays. Astronomers believe nearby supernovae, exploding stars, in the past 20 million years created the bubble and excited its remaining gas.

THE LOCAL BUBBLE

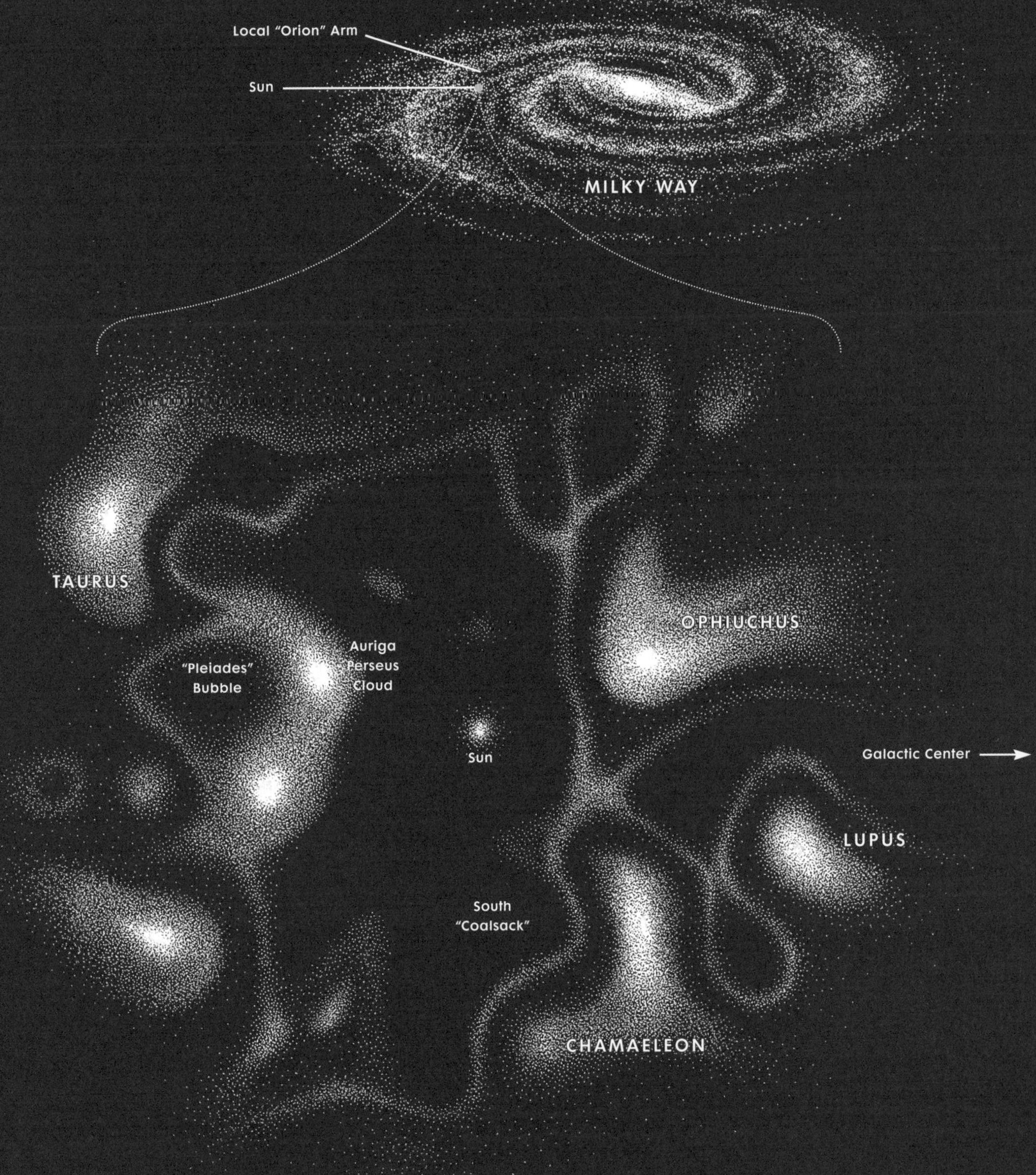

Local "Orion" Arm

Sun

MILKY WAY

TAURUS

OPHIUCHUS

"Pleiades"
Bubble

Auriga
Perseus
Cloud

Sun

Galactic Center ⟶

LUPUS

South
"Coalsack"

CHAMAELEON

THE MAGELLANIC BRIDGE

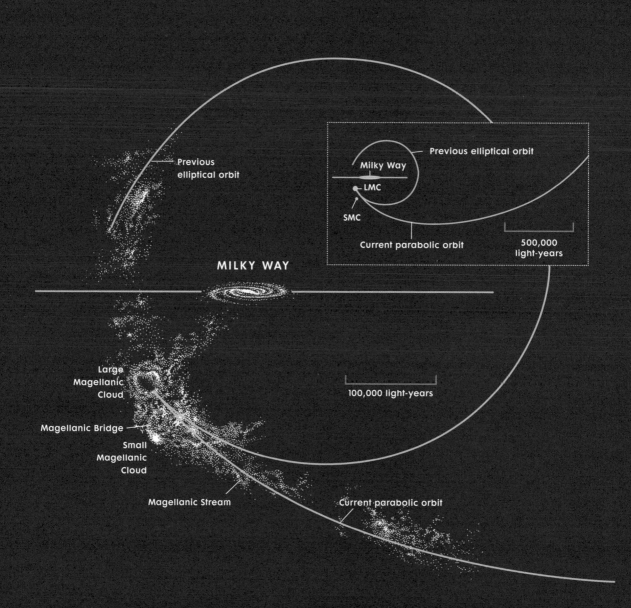

Previous elliptical orbit

Previous elliptical orbit

Milky Way

LMC

SMC

Current parabolic orbit

500,000 light-years

MILKY WAY

Large Magellanic Cloud

Magellanic Bridge

Small Magellanic Cloud

Magellanic Stream

Current parabolic orbit

100,000 light-years

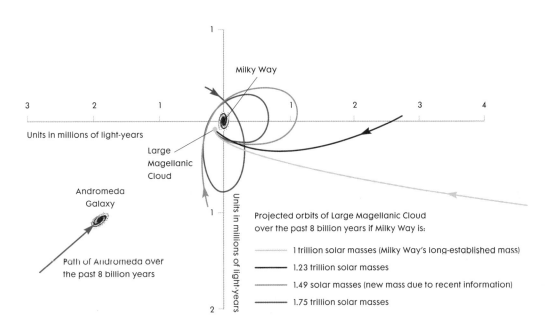

3 2 1 1 2 3 4

Units in millions of light-years

Milky Way

Large
Magellanic
Cloud

Andromeda
Galaxy

Path of Andromeda over
the past 8 billion years

Units in millions of light-years

1

2

Projected orbits of Large Magellanic Cloud
over the past 8 billion years if Milky Way is:

— 1 trillion solar masses (Milky Way's long-established mass)

— 1.23 trillion solar masses

— 1.49 solar masses (new mass due to recent information)

— 1.75 trillion solar masses

Above The orbits of the Large and Small Magellanic Clouds will determine where they end up after the Milky Way and Andromeda galaxies merge. The Clouds' orbits relate to the Milky Way's mass, which has been a recent area of debate. A higher mass implies a more tightly bound orbit of the Large Magellanic Cloud. If the Milky Way is 1.75 trillion solar masses, then the LMC likely will remain bound to the resultant galaxy. If, instead, the Milky Way is only 1 trillion solar masses, the LMC has had a more unbound orbit and thus likely will be tossed into intergalactic space during the merger. Whatever happens with the LMC, the Small Magellanic Cloud will likely follow suit.

Opposite **THE MAGELLANIC BRIDGE** Both of the Magellanic Clouds possess a lighter density fog of hydrogen surrounding them called the Magellanic Bridge, as well as a long tail of similar material that trails out behind them, known as the Magellanic Stream. This illustration displays their positions relative to each other and the Milky Way, as well as their recently recalculated parabolic path.

Quite a few other dwarf galaxies, containing small masses, lie relatively close to the Milky Way, and many of them are orbiting our galaxy. Among them are the Sagittarius Dwarf Galaxy, the Sculptor Dwarf, the Draco Dwarf, the Fornax Dwarf, the Ursa Minor Dwarf, and Leo I and Leo II. The gravitational pull of larger galaxies often tugs such tiny galaxies inward, and eventually the larger galaxies consume their smaller neighbors. This is happening right now with the Sagittarius Dwarf Spheroidal Galaxy, which is tidally distorted and being slowly shredded and incorporated into the Milky Way.

Can you imagine what Earth's night sky will look like when a large spiral galaxy approaches us? The form of the Andromeda Galaxy, now visible as a hazy smudge, will loom larger and larger in our sky as the galaxy flies toward Earth. Eventually, the view for

FAST FACT
All of the stars you see in the night sky with the unaided eye—as many as 2,000 at a perfectly dark site—belong to the Milky Way Galaxy. Only a few other, more distant, galaxies are visible to the eye alone—as fuzzy patches of light.

anyone alive in either galaxy will be incredible. Of course by the time this occurs, Earth itself will have long been inhospitable. But countless other worlds may well have beings that gaze upward at what would be, to us, a very strange sight. To them, it will be the norm, as motions in the universe take place over long intervals. At any given time, we get to see what amounts to one frame of a gigantic, incredibly long cosmic movie.

Yet another great pair of colliding galaxies can be seen in NGC 4676A and NGC 4676B in Coma Berenices, often called "the Mice." These two galaxies are an odd pair: one is a barred spiral galaxy and the other is an irregular galaxy; as astronomers often do with such distorted objects, their classifications are appended with the term "peculiar," to signify a measure of distortion in their form. The Mice lie an impressive 290 million light-years away, and yet they are bright enough to be observed and imaged by amateur astronomers.

Of the almost countless galaxies in the sky, astronomers know of thousands of interacting galaxies. Many of the brightest were cataloged in the 1960s by the American astronomer Halton C. Arp, who wrote the celebrated *Atlas of Peculiar Galaxies,* which was published in 1966. This seminal work offers 338 of the best examples of distorted galaxies, most of them caused by interactions. The catalog is still used as a source of information on many of the most intriguing objects in the sky.

The Future of the Milky Way:

HOW GALAXIES COLLIDE

In 2008, the American theoretical physicist Abraham Loeb and his colleagues at Harvard University determined that in several billion years the Andromeda Galaxy and the Milky Way will collide and begin to merge. In studying the nature of the dark matter halos, which will play an important role in the upcoming train wreck, Loeb and his colleagues created models of the collision between the two galaxies in a series of computer simulations. They dubbed the eventual supergalaxy that will take shape from the remnants of the Milky Way and Andromeda "Milkomeda."

The astronomers found that the two galaxies will make a close first pass less than 2 billion years from now. The collision and merger itself will take place on a time scale of something just shy of 5 billion years. At first, this encounter will simulate a slow dance in space—a set of waltzes to the eyes of any inhabitants of planets within both galaxies. As the encounter continues, the forms of the galaxies will draw closer and closer and then the merger will accelerate before being completed at the point where the dark matter halos of the two galaxies are separated by about 300,000 light-years.

THE MILKY WAY GALAXY RINGED BY AIRGLOW
A vertical slice of the Milky Way stands straight up in this striking image made in Australia. The ethereal rings of orangey light framing the galaxy's disk are caused by airglow, a process of chemiluminescence when oxygen and nitrogen atoms react with hydroxyl ions high in Earth's atmosphere. The effect is of a galaxy surrounded by a spooky halo of glowing light.

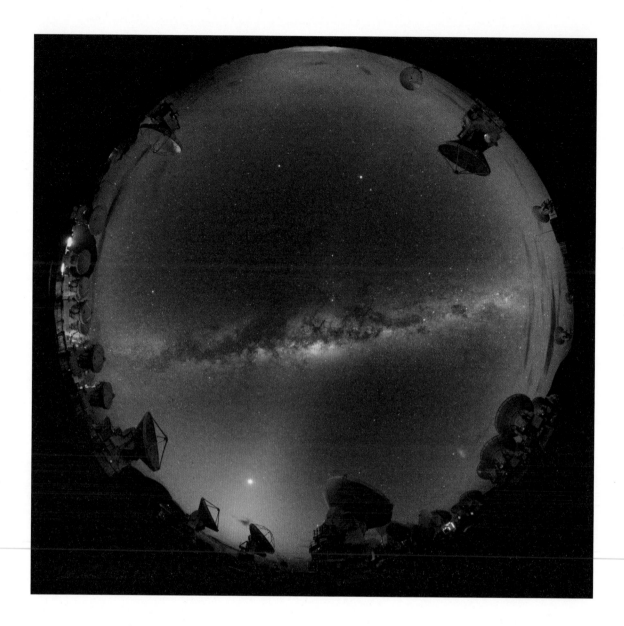

Previous THE MILKY WAY SHINES OVER THE ATACAMA DESERT

Our galaxy, the Milky Way, explodes with color in this image taken in Chile's high Atacama Desert, perhaps the darkest sky on Earth. The line of the Milky Way is peppered with bluish and pinkish light from glowing star clusters and nebulae, and the Small Magellanic Cloud appears at lower right.

Above **THE GHOSTLY GLOW OF THE MILKY WAY**

The disk of the Milky Way Galaxy stretches across our night sky, and it is visible overhead from a dark site during the right season and time of night. What we see is the unresolved light from billions of stars scattered along our galaxy's disk, as viewed from within. We can't know exactly what our galaxy looks like from outside, but only have an approximation based on mapping the Milky Way's structure. In this view, the radio dishes of Chile's ALMA Observatory stand in the foreground.

Opposite **PINK REGIONS OF STAR FORMATION LIGHT UP THE LMC**

The Large Magellanic Cloud glows in bright pinks and purples as imaged with the European Southern Observatory's 1-meter telescope. The largest star factory is the Tarantula Nebula, left of center.

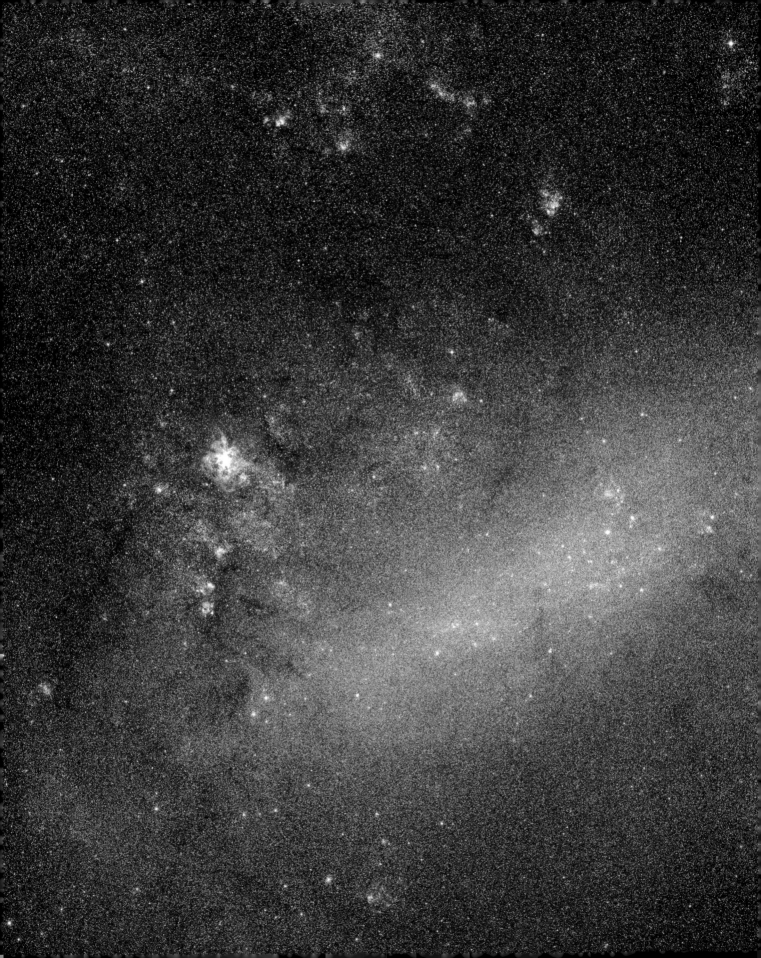

Previous **THE LARGE MAGELLANIC CLOUD IN ULTRAVIOLET LIGHT**
To capture the LMC in ultraviolet, NASA's Swift satellite made this multi-image mosaic. Recently formed stars in the Milky Way satellite appear prominent in this picture, and in the upper left is the gargantuan Tarantula Nebula, NGC 2070.

Above **THE MILKY WAY'S SATELLITE IN DRACO**
The Draco Dwarf is a dwarf spheroidal galaxy that orbits the Milky Way at a distance of some 260,000 light-years. It was discovered at Lowell Observatory in 1954. Recently, astronomers have found the galaxy has a very high concentration of dark matter.

Opposite **THE SMALL MAGELLANIC CLOUD AND 47 TUCANAE**
As imaged with the VISTA Telescope of the European Southern Observatory, the Small Magellanic Cloud is aglow with star forming regions, areas of pinkish light. To the right of the galaxy is the mammoth globular star cluster 47 Tucanae, one of the brightest in our sky.

THE FORNAX DWARF GALAXY: A MILKY WAY SATELLITE
The luminous fuzzball known as the Fornax Dwarf is
a dwarf elliptical galaxy discovered in 1938. It is a
satellite of the Milky Way, lying at a distance of 460,000
light-years, and contains a very bright globular star
cluster, NGC 1049, that is visible in backyard telescopes.

1

Pages 100-105 Most galaxies are moving apart from one another as the universe expands. But local motions and gravity mean that some are falling toward one another. Such is the case with the Milky Way and the Andromeda Galaxy, our nearest large galactic neighbor, which lies 2.5 million light-years away. Closing in at a velocity of 190 miles per second, Andromeda will merge with our galaxy some 4 billion years from now. This magnificently produced sequence shows our night sky as it appears today, with Andromeda to the left of the Milky Way (image 1), the growing size of Andromeda as it begins to approach (image 2), the chaotic, tangled arms of both galaxies 3.9 billion years from now (image 3), the glow of intermingled galaxies 5.1 billion years from now (image 4), and the full merger of the two galaxies, some 7 billion years from now (images 5 and 6).

2

4

5

6

ARP 269: INTERACTING GALAXIES IN CANES VENATICI
One of the better-known examples of galaxy interaction
lies at a distance of 25 million light-years in the
springtime constellation Canes Venatici. NGC 4485 and
NGC 4490, collectively known as Arp 269, are circling
each other and over time will merge. The brighter and
larger NGC 4490 is sometimes called the Cocoon Galaxy.

Chapter Three

NEARBY GALAXIES: THE LOCAL GROUP

✳ ✳ ✳

Astronomers now have a pretty good inventory of our cosmic neighborhood. We have already discussed some of the most prominent, including the Andromeda Galaxy and the Magellanic Clouds. At least fifty-four galaxies hang together in the Local Group, gravitationally bound as sisters in the same community. As with stars, it's hard to know the exact count: not only are the majority of these galaxies tiny dwarfs that are hard to detect unless they are quite close, but some as-yet-undiscovered dwarf galaxies that lie nearby could be obscured by our own Milky Way's plane of dust. But let's say, conservatively, that at least fifty-four galaxies belong to our clan (and likely more).

DISCOVERING OUR
GALACTIC BRETHREN

As we have seen, the universe is expanding everywhere we look. Hubble, Slipher, Humason, and others first explored the evidence for the expanding universe in the 1910s and 1920s. Since the 1960s, astronomers have accumulated ironclad evidence that the universe started in the Big Bang, some 13.8 billion years ago. As we have also seen, the expansion of the universe doesn't prevent some galaxies from hanging out close together, on local scales. Gravity sees to that. During the second half of the twentieth century, astronomers began to piece together the story of the nearest galaxies surrounding us, which came to be called the Local Group of galaxies. Edwin Hubble introduced the term in his 1936 work *The Realm of the Nebulae*, and the term stuck.

FAST FACT
When Hubble recognized the Local Group's existence, he knew of twelve member galaxies. That number has more than quadrupled since his discovery.

Think about our spaceship journey out into the cosmos. We've seen that traversing the Milky Way would take 100,000 years if we traveled at the speed of light. Moving outward at the same speed, traversing the Local Group from end to end, would take us 10 million years. Traveling from Earth to the Local Group galaxies would take several million years. The Andromeda Galaxy, our brightest and largest neighbor galaxy, is 2.5 million light-years away. The light that strikes your eye as you look at the Andromeda Galaxy in a telescope has been traveling in space for 2.5 million years. It's easy to visualize the size of our Local Group compared to our own Milky Way: the Local Group's diameter is about 100 times larger than our galaxy's disk. This mental picture, however, only begins to give us an appreciation of the vastness of the cosmos. We're still just dipping one toe into the cosmic ocean.

THE MEMBERS OF THE LOCAL GROUP

Astronomers believe that our Local Group is shaped like a spherical cloud about 10 million light-years across. The two areas within that sphere with the most mass, the "hot spots," are the areas surrounding the Andromeda Galaxy and the Milky Way. The third relatively large galaxy in the Local Group is the Triangulum, or

Pinwheel, Galaxy, M33. Aside from the "big three," the rest of the galaxies in the Local Group are small players. They are important to us as neighbors, however, because they allow astronomers to study a variety of galaxy types.

Coinciding with the realization of what galaxies are, the first members of the Local Group to be identified as galaxies were, of course, the Milky Way and the Andromeda Galaxy. Understanding the Magellanic Clouds soon followed. A short time later, astronomers realized that M33 was a nearby member of our Local Group. The fainter members have been found and identified as nearby galaxies more recently, and some just within the last generation.

THE SATELLITES OF THE MILKY WAY

Besides the bright Magellanic Clouds, a retinue of at least twelve other satellite galaxies orbit the Milky Way Galaxy. We'll look at them in terms of increasing distance.

In 2003, astronomers found evidence for what could be the closest dwarf galaxy orbiting the Milky Way, the Canis Major Dwarf. Although its existence is still not entirely agreed upon, this galaxy probably has something like the mass of a billion suns, is believed to lie 25,000 light-years away, and is being shredded by the Milky Way's gravitational forces. The structure was detected using an infrared survey called 2MASS, and so the interpretation of the data is not entirely clear, as this area is heavily obscured. Astronomers may simply be seeing peculiar outlying stars of the Milky Way itself.

More certain to be a Milky Way satellite is the Sagittarius Dwarf Galaxy, which turned up in a survey in 1994. As with all dwarf galaxies, it is named for the constellation within which it lies in our sky. It is shaped elliptically, is only about 65,000 light-years away, and stretches 10,000 light-years across. It contains a small amount of mass and is accompanied by four globular clusters (the spherical balls of old stars), one of which, M54, is familiar to backyard observers and lies near the galaxy's center. Astronomers have calculated that the Sagittarius Dwarf Galaxy has

GALACTIC PLANE

62 MILLION LIGHT-YEARS

Sun

As the Sun and solar system orbit the galactic center, we bob up and down like a carousel horse. Thus, every 62 million years, we enter a period in the northern part of the galaxy's disk, an area with more exposure to cosmic rays, energetic particles that bombard us from space. This could be related to past extinction events on Earth.

plunged through the plane of the Milky Way several times in its elliptical, looping orbits around us. The galaxy will again plunge through the Milky Way's disk, stripping it of more mass, about 100 million years from now. Ultimately, the Sagittarius Dwarf Galaxy will slow down and be absorbed into the Milky Way, another victim of galactic cannibalism.

Next in line in terms of distance is the Ursa Major II Dwarf Galaxy, a dwarf spheroidal galaxy some 100,000 light-years away. Discovered in 2006, it is similar to other dwarf satellite galaxies in containing old stars that formed mostly 10 billion years ago or more.

Other dwarf galaxies orbiting the Milky Way offer astronomers a varied set of strange, small galaxies to explore with modern science. The Ursa Minor Dwarf Galaxy was discovered by astronomers who made a systematic sky survey at California's Palomar Observatory in 1955. It lies about 200,000 light-years away and, characteristic of these systems, seems to have long ago ceased its active star formation.

THE LOCAL GROUP

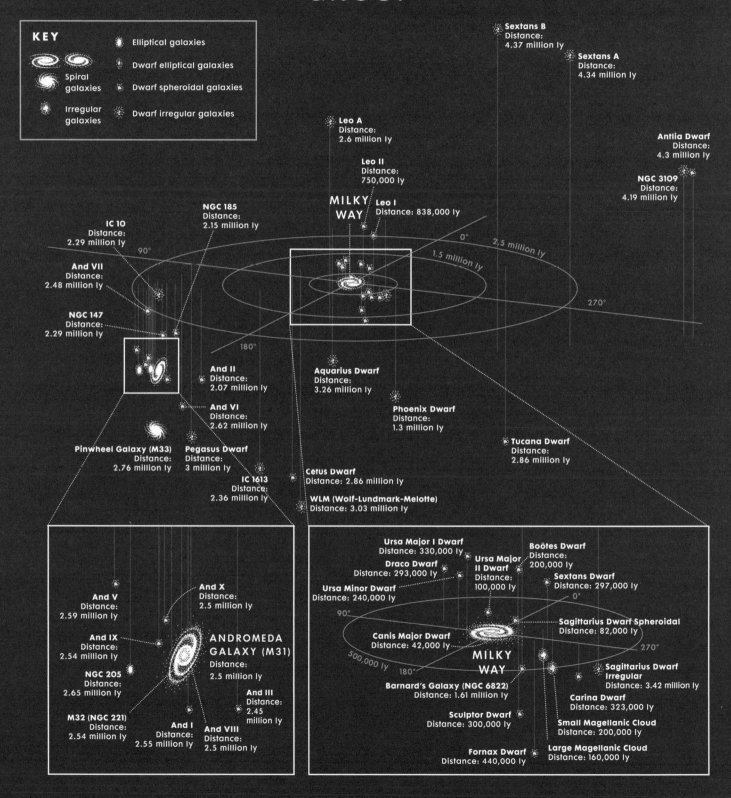

KEY

- Elliptical galaxies
- Dwarf elliptical galaxies
- Dwarf spheroidal galaxies
- Dwarf irregular galaxies
- Spiral galaxies
- Irregular galaxies

Sextans B
Distance:
4.37 million ly

Sextans A
Distance:
4.34 million ly

Leo A
Distance:
2.6 million ly

Antlia Dwarf
Distance:
4.3 million ly

NGC 3109
Distance:
4.19 million ly

Leo II
Distance:
750,000 ly

MILKY WAY

Leo I
Distance: 838,000 ly

NGC 185
Distance:
2.15 million ly

IC 10
Distance:
2.29 million ly

And VII
Distance:
2.48 million ly

NGC 147
Distance:
2.29 million ly

0°

2.5 million ly

1.5 million ly

90°

270°

180°

And II
Distance:
2.07 million ly

Aquarius Dwarf
Distance:
3.26 million ly

And VI
Distance:
2.62 million ly

Phoenix Dwarf
Distance:
1.3 million ly

Pinwheel Galaxy (M33)
Distance:
2.76 million ly

Pegasus Dwarf
Distance:
3 million ly

IC 1613
Distance:
2.36 million ly

Cetus Dwarf
Distance: 2.86 million ly

Tucana Dwarf
Distance:
2.86 million ly

WLM (Wolf–Lundmark–Melotte)
Distance: 3.03 million ly

And V
Distance:
2.59 million ly

And X
Distance:
2.5 million ly

And IX
Distance:
2.54 million ly

ANDROMEDA GALAXY (M31)
Distance:
2.5 million ly

NGC 205
Distance:
2.65 million ly

And III
Distance:
2.45 million ly

M32 (NGC 221)
Distance:
2.54 million ly

And I
Distance:
2.55 million ly

And VIII
Distance:
2.5 million ly

Ursa Major I Dwarf
Distance: 330,000 ly

Boötes Dwarf
Distance:
200,000 ly

Draco Dwarf
Distance: 293,000 ly

Ursa Major II Dwarf
Distance:
100,000 ly

Sextans Dwarf
Distance: 297,000 ly

Ursa Minor Dwarf
Distance: 240,000 ly

0°

Sagittarius Dwarf Spheroidal
Distance: 82,000 ly

90°

270°

Canis Major Dwarf
Distance: 42,000 ly

MILKY WAY

Sagittarius Dwarf Irregular
Distance: 3.42 million ly

500,000 ly

180°

Barnard's Galaxy (NGC 6822)
Distance: 1.61 million ly

Carina Dwarf
Distance: 323,000 ly

Sculptor Dwarf
Distance: 300,000 ly

Small Magellanic Cloud
Distance: 200,000 ly

Fornax Dwarf
Distance: 440,000 ly

Large Magellanic Cloud
Distance: 160,000 ly

The Draco Dwarf Galaxy also turned up during the Palomar Observatory Sky Survey. It lies some 260,000 light-years away and is about 2,500 light-years across, holding large amounts of dark matter.

The Sculptor Dwarf Galaxy was discovered by Harlow Shapley on plates that the astronomer, now at Harvard, took in 1937. Lying some 290,000 light-years away, this dwarf is an elliptically shaped cloud of stars.

The Sextans Dwarf Galaxy evaded astronomers until 1990, when they detected it some 300,000 light-years away. This cloud of stars stretches about 8,400 light-years across.

The galactic zoo of Milky Way companion galaxies goes on. In 1977, astronomers found the Carina Dwarf Galaxy, a small companion galaxy that lies some 330,000 light-years away, measures only one-seventy-fifth of the diameter of the Milky Way, and consists of a cloud of stars that is heavily disrupted by the Milky Way's gravitational forces.

The Ursa Major I Dwarf Galaxy lies some 330,000 light-years away, spans a few thousand light-years across, and was discovered only in 2005.

Harlow Shapley also discovered the Fornax Dwarf Galaxy, another elliptical cloud of stars, this one lying 460,000 light-years away and notable for its group of six globular star clusters. The brightest of these globulars, NGC 1049, is visible in backyard telescopes and was known long before the Fornax Dwarf Galaxy itself.

Two distant Milky Way satellites lie in the constellation Leo. Leo II is a dwarf spheroidal galaxy that lies about 690,000 light-years away, making it one of the more distant satellites of the Milky Way. Discovered in 1950 on the Palomar Sky Survey plates, it is believed to hold about 2.7 billion solar masses and extends 1,500 light-years across.

Opposite **MAPPING THE LOCAL GROUP OF GALAXIES**
In 1936, American astronomer Edwin Hubble introduced the concept that our Milky Way was part of a small group of galaxies he called the Local Group. He listed its members as the Milky Way, the Large and Small Magellanic Clouds, the Andromeda Galaxy (M31), M32, the Triangulum Galaxy (M33), NGC 147, NGC 185, NGC 205, NGC 6822, and IC 1613, plus IC 10 as a possible member. That number has grown to 54 current members, of which 33 are satellites of M31 and 14 are satellites of the Milky Way.

Leo I is a curious member of the Local Group in that this dwarf spheroidal galaxy lies a short distance in our sky from one of the northern sky's prominent stars, Regulus, the brightest star in Leo. So Leo's position is easy to find, but the galaxy is so faint that observing it requires a large telescope. Leo I lies 820,000 light-years away, was discovered in 1950 during the Palomar Sky Survey, and holds some 20 million solar masses of material.

These dwarf galaxies form the bulk of the satellites orbiting our Milky Way, which comprises some

24 galaxies altogether. A few galaxies, including Maffei 1 and Maffei 2, were thought to be Local Group members for years but now are believed to lie somewhat beyond the gravitational grasp of our galactic neighborhood.

THE ANDROMEDA GALAXY

Beyond the Milky Way and its satellites, the big galaxy on the block in our cosmic neighborhood is the Andromeda Galaxy. The object on which Hubble's attention focused, Andromeda is a sprawling object, holding perhaps a trillion stars—close to twice as many as the Milky Way. Its bright disk stretches 220,000 light-years across, a distance about twice the diameter of our galaxy's disk. (Recent research suggests that the Milky Way's disk may be larger than 100,000 light-years, but this work needs more examination and confirmation.) The Andromeda Galaxy is so vast that, at a distance of 2.5 million light-years, we can see it, unaided by binoculars or telescopes, as a fuzzy patch of light in our night sky. Known as M31 in Charles Messier's famous catalog of deep-sky objects, Andromeda contains a whopping 1.5 trillion solar masses.

The Andromeda Galaxy formed approximately 10 billion years ago from the collision and coming together of numerous smaller galaxies. During this early time, the rate of star formation in M31 was high; it has fallen off to much lower levels more recently. Several billion years ago, the Andromeda and Triangulum Galaxies passed relatively close to each other; that passage again kicked up star formation for a time and disturbed the Triangulum Galaxy's disk.

Just as with the Milky Way, the Andromeda Galaxy has a large halo of hot gas and an even larger halo of dark matter. Andromeda has more old stars than the Milky Way and a somewhat higher luminosity, or absolute brightness, than our own galaxy. The current rate of star formation is only about one-third that of the Milky Way: Andromeda produces an equivalent of just one solar mass of new stars per year. The galaxy rotates at its highest speed of 160 miles per second (250 kilometers per second)—some 33,000 light-years out from the core—and contains a super-massive black hole in its center of about 100 million solar masses.

The Andromeda Galaxy formed approximately 10 billion years ago.

The Andromeda Galaxy contains quite a few HII regions—star-forming clouds that the German astronomer Walter Baade, who studied them, dubbed "pearls on a string." They dot the galaxy's spiral arms with pinkish glows and the blue-white hues of young stars. The galaxy contains a large star cloud (a rich concentration of stars) so striking that it earned its own designation in the *New General Catalog*, NGC 206.

THE SATELLITES OF
THE ANDROMEDA GALAXY

The Milky Way isn't the only Local Group galaxy with a cloud of satellite hangers-on. The Andromeda Galaxy holds at least nineteen significant dwarf galaxies in its gravitational clutch. Let's look at them, moving from the closest to us through the most distant.

The closest such satellite is NGC 185, discovered by the German-English astronomer William Herschel in 1787, which lies 2 million light-years away and also contains old stars, typical of dwarf spheroidal galaxies. Astronomers have found that there has been only minor star formation near the center of NGC 185 in the last billion years.

Next in line comes a trio of very small dwarfs. These include Andromeda II, discovered in 1970 and lying 2.2 million light-years away; Andromeda I (1970, 2.4 million light-years); and Andromeda III (1970, 2.4 million light-years).

The next galaxy is visible right along with the Andromeda Galaxy itself in a low-power telescope. It is M32, a dwarf elliptical galaxy that appears as a roundish puffball of light. M32 is bright enough to have been discovered by the French astronomer Guillaume Le Gentil in 1749. It lies essentially the same distance as the Andromeda Galaxy from us, 2.5 million light-years, but is just on the near side of the big galaxy. It spans about 6,500 light-years across and contains mostly old yellow and red stars. If you look carefully at images of the Andromeda Galaxy, you'll see a slight warping in the big galaxy's disk. Astronomers believe that this was caused by an off-center impact on the disk by M32 about 800 million years ago. M32 has a central black hole of about the same mass range as ours in the Milky Way.

FAST FACT

M32, the most luminous of Andromeda's satellites, has almost no cool gas and no stars younger than a few billion years. It may be the remnant center of what was a much larger galaxy.

Slightly more distant is the dwarf NGC 147; discovered by the English astronomer John Herschel in 1829, it lies 2.5 million light-years away. This dwarf spheroidal galaxy contains mostly old stars, its star-forming activity having ended about 3 billion years ago.

Next in line come more of the extremely faint dwarfs discovered in recent times. They are Andromeda V, discovered in 1998 and 2.5 million light-years away; Andromeda IX (2004, 2.5 million light-years); Andromeda VII (1998, 2.6 million light-years); and Andromeda XI (2006, 2.6 million light-years).

Slightly farther away, at a distance of 2.7 million light-years, is the other dwarf galaxy that can be observed in backyard scopes along with M31. Compared with M32, NGC 205 (sometimes called M110) is a somewhat fainter, more elongated dwarf galaxy that lies slightly farther from the center of the big spiral. In 1773, the great French comet hunter Charles Messier observed NGC 205. Messier compiled his famous catalog of fuzzy nebulae in order to avoid them when comet hunting, because comets and nebulae look similar in telescopes. But although he observed it, Messier never included NGC 205 in his list. So it has no formal Messier number, although later historians have sometimes stapled it onto the list, calling it M110. The German observer Caroline Herschel independently discovered this galaxy in 1783. The distance to NGC 205 is about 2.7 million light-years, meaning it is behind the Andromeda Galaxy as we see it. This is an unusual dwarf elliptical galaxy in that it contains a fair amount of dust and hints at recent star formation.

> Messier compiled his famous catalog of fuzzy nebulae in order to avoid them when comet hunting.

The next-in-line galaxies by distance make up another slew of extremely faint dwarfs discovered in recent times. They are Andromeda VI (discovered in 1999, 2.7 million light-years away); Andromeda VIII (2003, 2.7 million light-years); Andromeda XXI (2009, 2.8 million light-years); Andromeda X (2005, 2.9 million light-years); and Andromeda XXII (2009, 3.0 million light-years).

Further, the Andromeda Galaxy has several satellites of unknown or poorly known distances: the Pegasus Dwarf Spheroidal Galaxy, the Cassiopeia Dwarf Galaxy, Andromeda XIX (all discovered in 2009), and several other extremely faint dwarfs that have yet to be studied thoroughly.

THE ANDROMEDA GALAXY

THE TRIANGULUM GALAXY

THE TRIANGULUM GALAXY

The third big galaxy in the Local Group, the Triangulum Galaxy, M33, is sometimes cited as the faintest galaxy visible to the naked eye alone under a dark sky. (Expert observers, however, under fantastically dark skies, have reported seeing the even fainter M81.) This spiral is oriented slightly closer to face-on to us than Andromeda, so it's a little easier to see its internal features. The galaxy contains a variety of pinkish, star-forming regions, the largest and brightest of which, NGC 604, is visible through backyard telescopes.

M33 has a diameter of about 60,000 light-years and holds approximately 40 billion stars, about one-tenth the number found in the Milky Way. The galaxy lies about 2.7 million light-years away, a trifle more distant than the Andromeda Galaxy. The rate of star formation in M33 is a good deal higher than that in Andromeda. M33 contains a notable stellar mass black hole, one "weighing" 15.7 times the mass of the Sun, but it is unusual in that it contains no supermassive black hole. Most galaxies larger than dwarfs do harbor such a black hole, and so M33 is often held out as a notable exception to the rule.

The Triangulum Galaxy is the smallest spiral galaxy in the Local Group and may well orbit the Andromeda Galaxy. It contains a modest number of globular clusters, 54, as opposed to the Milky Way's approximately 150.

Astronomers debate whether the Triangulum Galaxy, M33, is itself a companion to the Andromeda Galaxy. The Pisces Dwarf Galaxy, a dwarf irregular lying 2.5 million light-years away, may be a satellite of the Triangulum Galaxy. Discovered in 1976, this odd galaxy is approaching the Milky Way at 180 miles per second (290 kilometers per second). Typically for such galaxies, it contains old stars and shows little evidence of star formation in at least the last 100 million years.

The Triangulum Galaxy is sometimes called the faintest galaxy visible to the naked eye.

OTHER LOCAL GROUP GALAXIES

The remaining galaxies in the Local Group are gravitationally bound to the group, but not to the three primary subgroups of Andromeda, Triangulum, and the Milky Way. They offer a fascinating landscape of objects, many of which are visible in backyard telescopes and are well known to amateur astronomers. Let's examine them, starting with the closest and then moving outward in space.

The Phoenix Dwarf is a dwarf irregular galaxy that was discovered in 1976 and lies 1.4 million light-years away. Thought at first to be simply a globular star cluster, the galaxy is a small object.

Barnard's Galaxy, cataloged as NGC 6822 and discovered in 1884 by the American astronomer E. E. Barnard, is one of the few galaxies visible in the constellation Sagittarius. It lies some 1.6 million light-years away and is classified as a barred irregular galaxy. Structurally, this galaxy is similar to the Small Magellanic Cloud.

IC 10 is an irregular galaxy in Cassiopeia lying 2.2 million light-years away. The American astronomer Lewis Swift discovered this strange object in 1887, and it was recently found to be a "starburst galaxy"—that is, it shows a recent burst of star formation, which is unusual for a small irregular galaxy.

The weird irregular galaxy IC 1613 lies in the constellation Cetus and is 2.4 million light-years away. Discovered by the German astronomer Max Wolf in 1906, IC 1613 is a strange, mottled galaxy that consists mostly of old stars, but like the Magellanic Clouds, it also has large, pinkish HII regions with ongoing star formation.

Discovered in 1999, the Cetus Dwarf is a dwarf spheroidal galaxy lying 2.5 million light-years away and containing an old population of red giant stars.

A similarly small irregular galaxy, Leo A, lies 2.6 million light-years distant and was found by the Swiss-American astronomer Fritz Zwicky in 1942. It contains some 80 million solar masses and its star formation seems to have become inactive long ago.

The irregular galaxy known as Wolf-Lundmark-Melotte was discovered by Max Wolf in 1909, and the astronomers Knut Lundmark and Philibert Melotte recognized it as a galaxy in 1926. It lies 3 million light-years away, and its star formation days are long over.

FAST FACT

The Local Group contains at least 54 galaxies, and likely as many as 100, as tiny dwarf galaxies are difficult to detect and many more may be discovered over time.

The Aquarius Dwarf Galaxy is a dwarf irregular. Found by astronomers in 1959, it lies at a distance of 3.2 million light-years away.

The Tucana Dwarf, a dwarf spheroidal galaxy discovered in 1990, contains old stars and lies at a distance of 3.2 million light-years.

The Sagittarius Dwarf Irregular Galaxy, not to be confused with the Milky Way satellite (Sagittarius Dwarf Elliptical Galaxy), is a dwarf about 3.4 million light-years away that was discovered in 1977.

Once thought to be a small cloud of a few galaxies, the Local Group is now known to contain dozens of galaxies, most of them dwarf irregulars or dwarf spheroidals. The "big three," the Andromeda Galaxy, the Milky Way, and the Triangulum Galaxy, offer astronomers a vast laboratory in which to study how galaxies work, both from the inside and the outside. The Local Group's many small galaxies are at least as important: they are close enough for astronomers to study in high resolution, both with telescopes in orbit, such as the Hubble Space Telescope, and with large ground-based instruments. They show the behavior of stars and other objects that would be too faint for astronomers to study at larger distances, outside the Local Group. A significant amount of our knowledge of how stars behave comes from studying the Magellanic Clouds, for example. Other, unusual galaxies in the Local Group, like Barnard's Galaxy in Sagittarius, offer laboratories for studying how galaxies evolve.

Some astronomers liken galaxies in the universe to buildings in cities. Certainly the large skyscrapers, the enormous office buildings, tell us a lot about how cities work. But the dwarfs, the ordinary little houses spread here and there, tell a critically important part of the story too. They build a foundation for astronomers' understanding of how galaxies work as systems that is vitally important before they move progressively farther out into the cosmos.

INSIDE THE LOCAL GROUP: HOW BLACK HOLES POWER GALAXIES

Back in the 1960s, astronomers had more problems to worry about than simply trying to decipher the extent of the universe. In 1963, the Cal Tech astronomer Maarten Schmidt made a very strange observation. He was studying an object, designated as 3C 273, that had been discovered in the 1950s in radio waves. The object appeared "loud" in radio noise, but in terms of optical light on photographic plates, it was just a faint smudge.

With very careful measurements, Schmidt was able to record a spectrum of the object and calculate its distance based on the redshift. Amazingly, Schmidt found that the object was 2.4 billion light-years away—a very distant object! Because it looked like a fuzzy star, he called 3C 273 a quasi-stellar object, or "quasar" for short. Many other similar objects were discovered later, such as 3C 321, Markarian 509, and A2261-BCG.

Quasars offered a major mystery to astronomers. The quasar 3C 273 was so distant and yet also blasted out a great deal of energy. It also changed brightness over short time scales. How could something so far away be emitting so much energy?

OTHER HIGH-ENERGY GALAXIES

Over the coming decade, astronomers turned up more highly energetic, distant objects that were blaringly loud in radio waves, gamma rays, and ultraviolet energy. In addition to quasars like 3C 273, there were Seyfert galaxies—named for Carl Seyfert, the American astronomer who identified them—which seemed to be lower-energy versions of quasars. And there were BL Lacertae objects, named for the first discovered example, originally thought to be a variable star in the constellation Lacerta. Nicknamed blazars, BL Lacertae objects radically changed brightness over even shorter time scales than quasars. Astronomers discovered more and more mysterious high-energy objects, including Hercules A, Markarian 231, and Cygnus A. They suspected that all of these objects were extremely distant, high-energy galaxies, but no one knew for sure. And slowly an even older mystery crept into the picture.

THE ORIGIN OF BLACK HOLES

In 1783, the English natural philosopher John Michell suggested that "dark stars" might exist, regions where gravity was so strong that nothing, not even light, could escape from them. His idea of what came to be called "black holes" languished for nearly 200 years for lack of any means of finding evidence of them, though Albert Einstein's theories of the early twentieth century had also predicted them. In a 1939 paper on black holes, Einstein worried that they should be around and wondered why they had not been found. Finally, in the 1970s, astronomers started gathering evidence that such regions might actually exist, and the discovery of objects like 3C 273 and the Circinus Galaxy added to astronomers' understanding.

Astronomers had long believed, based on theoretical calculations, that the easiest way to form a star-sized black hole was through the death of a massive sun, which would collapse gravitationally after it could no longer undergo nuclear fusion reactions. They predicted that any star with more than about five times the mass of our Sun would end up as a black hole. By the 1970s, astronomers had several collapsed-star black hole candidates. The leading one was Cygnus X-1, a binary star located 6,000 light-years away and first recorded as a strong X-ray source. Soon the story of Cygnus X-1 would also be tied to galaxies like M77, M82, and NGC 4725.

THE FAMOUS BLACK HOLE BET

In 1975, the astrophysicists Kip Thorne of Cal Tech and his friend Stephen Hawking of Cambridge University famously made a bet about whether Cygnus X-1 would be shown to be a black hole. By 1990, virtually all astronomers had agreed that the only explanation for the extremely high velocities observed in Cygnus X-1's accretion disk was that a black hole, caused by a collapsed star, resided at its center. Fifteen years had passed, and Thorne won the bet—and a year's subscription to *Penthouse*, for which he had wagered. (Hawking had wagered for a year's subscription to *Popular Mechanics*.)

> In 1975 Stephen Hawking and Kip Thorne famously bet about whether Cygnus X-1 would be shown to be a black hole.

As astronomers became comfortable with the first confirmed black holes like Cygnus X-1, they also continued piling up more and more observations of high-energy objects like 3C 273. Eventually, by the 1990s, they realized that what they had once believed were lots of different objects—quasars, Seyfert galaxies, and BL Lacertae objects—were mostly the same. The apparent differences had largely resulted from viewing them from different angles. They began to call these objects active galaxies, or active galactic nuclei (AGN), as the blasts of high energy appeared to come from the galaxies' centers. But still the mystery remained: what could be producing the enormous amounts of radiation shooting away from the centers of these galaxies?

THE PICTURE OF BLACK HOLES COMES TOGETHER

The timing on black holes couldn't have been better. As astronomers observed more and more active galaxies, they theorized that another type of black hole—black holes in the centers of galaxies—might exist. These would be the same creatures as Cygnus X-1—regions of gravity collapsed so completely that light and everything else was trapped in space-time. But rather than containing the mass of at least five stars like the Sun, these supermassive black holes would contain the mass of millions of stars like the Sun. Many unusual, high-energy galaxies showed signs of internal black holes, like NGC 2276, and other, more exotic objects, like Einstein's Cross, also demonstrated the importance of understanding Einstein's theory of relativity.

At first, the evidence for supermassive black holes in the centers of galaxies was tough to come by. But in 1988 two teams of astronomers published studies of the Andromeda Galaxy, our sister spiral in the Local Group. Led by John Kormendy of the University of Texas and by Alan Dressler of the Carnegie Institution for Science and Douglas Richstone of the University of Michigan, the two teams used ground-based telescopes that observed gas clouds in the galaxy's center rotating at incredibly rapid speeds. These could only be explained if the center contained millions of times the mass of the Sun, concentrated in a tiny region only about the size of the solar system. And the only way to produce such a region of space known to astrophysics was with a supermassive black hole. Galaxies like 0313-192 would help to prove this.

SUPERMASSIVE BLACK HOLES

Soon thereafter, astronomers identified central supermassive black holes via similar observations in the Sombrero Galaxy (M104) in the constellation Virgo and in NGC 3115 in Sextans. And then came discoveries of a central supermassive black hole in the galaxy M106 in Canes Venatici and in our own Milky Way. Follow-up observations with the Hubble Space Telescope began to paint a picture of central supermassive black holes being common in galaxies. That would turn out to be an understatement.

Astronomers now believe that most normal galaxies have a central supermassive black hole, except for dwarf galaxies, which typically lack the mass to have formed a black hole in the first place. Recent studies have suggested that our Milky Way's black hole "weighs" between 3.7 and 4.3 million suns. But there are exceptions: in our own Local Group, the Pinwheel Galaxy (M33), a nicely sized spiral, shows no evidence of a central supermassive black hole. And as yet, no one knows why.

As the flood of evidence surrounding central supermassive black holes poured in, astronomers came to realize that the active galaxies observed since the 1960s, like 3C 273, were young galaxies in the very early universe, being powered by central supermassive black holes. The central black hole itself does not emit the energy seen

> Think about a black hole as a near-relativistic-speed version of water swirling around your bathtub drain.

———

coming from it; it is *black*, after all, and swallowing up everything it can. But gas, dust, and stars heat up more and more as they slingshot around the black hole, accelerating to incredibly high speeds. As they heat up, they emit an incredible blaze of radiation that we can see from very far away. We see this in the centers of galaxies like M77 and M106. Think of this as a near-relativistic-speed version of water swirling around your bathtub drain, some of it circling faster and faster without falling into the drain hole. In some active galaxies, dramatic outflows of energy shoot out in the form of long jets. The light produced from active galaxies can be a thousand times brighter than our entire Milky Way Galaxy.

WHEN GALAXIES WERE YOUNG

The Milky Way and other galaxies with central supermassive black holes would have once behaved similarly when younger, and they could enter periods of highly active behavior again when gas, dust, and stars make their way toward their centers in the future. In fact, the Milky Way's black hole only came into focus in 2002 when astronomers began publishing data showing rapid motions of stars and gas clouds around a bright, compact object, dubbed Sagittarius A* (pronounced "Sagittarius A-star"), in our galactic center. But rather than finding it to be a star, they determined that it was the effect of the galaxy's central supermassive black hole accelerating and heating up the surrounding stars and gas clouds. Yet Sagittarius A*'s brightness is significantly dimmer than that of an active galaxy.

In 2002, astronomers also detected a gas cloud, dubbed G2, that seemed to be falling toward the black hole at the center of the Milky Way. In 2014, they expected to see this gas cloud absorbed and our own galaxy becoming briefly active with flares of energy. However, when the time came, this did not happen, suggesting that G2 is not a simple gas cloud but may contain a central star. Or perhaps it's a dense region of a streamer of material that is more extensive than the part we can now see.

Opposite STAR FORMING REGIONS LITTER THE TRIANGULUM GALAXY

M33, known as the Triangulum Galaxy, contains a vast storehouse of pinkish HII regions, areas of active star formation. The galaxy's innermost 30,000 light-years are shown in this dazzling view, capturing some of the largest stellar nurseries known in the cosmos. Intense radiation from newborn blue-white stars ionizes the surrounding hydrogen gas, setting it aglow in pink.

Above **THE ANDROMEDA GALAXY IN INFRARED LIGHT**

An astounding color image of the Andromeda Galaxy combines visible light and infrared data from the Spitzer Space Telescope. The infrared light, shown in red and green, reveals lumpy dust lanes warmed by young stars as they wrap tightly toward the galaxy's core. The companion galaxies M32 (above center) and NGC 205 (bottom right) are also visible.

Above **LEO I: A DWARF GALAXY BESIDE THE BRIGHT STAR REGULUS**

Local Group dwarf Leo I is easy to find on the sky, as it is adjacent to Leo's brightest star, Regulus. It is faint, however. This dwarf lies 820,000 light-years away and is one of the most distant satellites of the Milky Way.

Opposite **CORE OF THE ANDROMEDA GALAXY**

Our closest big galactic neighbor, the Andromeda Galaxy, lies some 2.5 million light-years away and has a bright central hub and core. Deep within the nucleus lies a supermassive black hole with the mass of 100 million solar masses.

A CLOSE-UP LOOK AT THE ANDROMEDA GALAXY'S ARMS
This spectacular Hubble Space Telescope view shows a slice of the Andromeda Galaxy's spiral arms that includes a whopping star cloud, a region of active star formation, just above center. It is bright enough to have its own designation, NGC 206, and spans an amazing 4,000 light-years.

Above **ANDROMEDA COMPANION M32 SHOWS HOT BLUE STARS**
The elliptical galaxy M32, one of the satellites of the Andromeda Galaxy, shows a population of hot, blue stars in this image made with the Hubble Space Telescope. The picture reveals a glow in ultraviolet light coming from hot, helium-burning stars late in their lives.

Opposite **BARNARD'S GALAXY: A BARRED IRREGULAR**
The barred irregular galaxy NGC 6822, called Barnard's Galaxy, lies 1.6 million light-years away in Sagittarius. It was discovered by Edward E. Barnard in 1884 and is a popular and somewhat challenging target for backyard galaxy hunters.

IC 10: A LOCAL GROUP IRREGULAR GALAXY
Lying at a distance of 2.2 million light-years
in the constellation Cassiopeia, IC 10 was
discovered by astronomer Lewis Swift in
1887 and identified as a galaxy in 1935. This
object is the only starburst galaxy in the
Local Group.

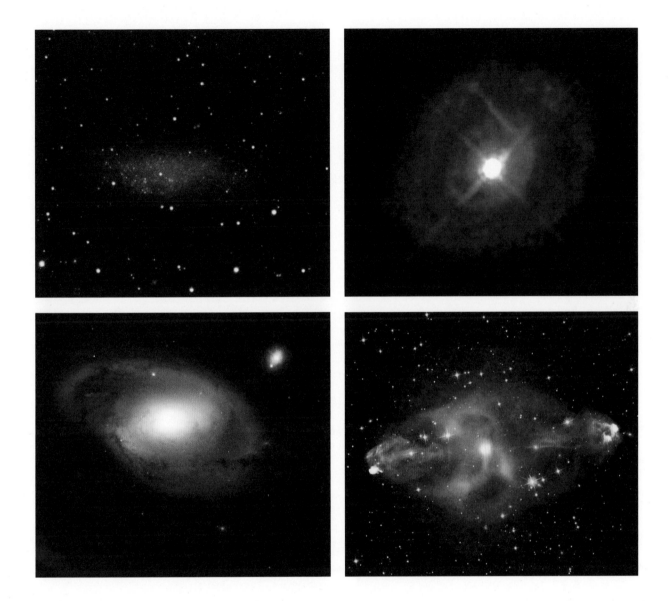

Top Left **WLM: A STRANGE IRREGULAR LOCAL GROUP GALAXY**
The odd irregular galaxy WLM, short for Wolf-Lundmark-Melotte, lies
3 million light-years away in the constellation Cetus. It was discovered
in 1909 by Max Wolf, and confirmed as a galaxy by astronomers Knut
Lundmark and Philibert Melotte in 1926.

Bottom Left **A FAMOUS PAIR OF CONTROVERSIAL OBJECTS**
During the 1970s astronomer Halton C. Arp and his colleagues claimed
to have imaged a light bridge between galaxy NGC 4319 and the tiny
quasar Markarian 205 (top right). This would have upset the apple
cart on cosmic distance measurements: redshifts indicated the two
were widely separated. In the end, Arp was wrong and damaged his
reputation. NGC 4319 lies 77 million light-years away, in Draco, and
Markarian 205 far beyond it, some 1.1 billion light-years away.

Top Right **MARKARIAN 509: A TURBULENT BLACK HOLE'S
INNER SECRETS**
The remote galaxy Markarian 509 lies at a distance of 500 million
light-years in the constellation Aquarius. Astronomers have found a hot
corona of gas surrounding the galaxy's inner region, along with cold
gaseous "bullets" flying outward at 1 million miles per hour. This explosive
core seems to be the result of the inner black hole, which acts as a
powerful central engine.

Bottom Right **CYGNUS A: A RADIO GALAXY WITH
SPECTACULAR LOBED JETS**
Named strangely because it is a radio source, Cygnus A is a powerful
galaxy lying some 600 million light-years away. This multiwavelength
image shows X-ray emission around the galaxy's center, in blue, and
radio energy in red. The galaxy's radio jets extend some 300,000 light-
years across and have been ejected by the central black hole.

Previous BLACK-HOLE-POWERED JETS FLANK HERCULES A

The enormous elliptical galaxy Hercules A contains a huge central black hole, "weighing in" at some 4 billion solar masses. Material falling in toward the black hole but not inside it gets whipped around and shot outward in polar jets, visible in this image. The galaxy lies some 2.1 billion light-years away.

Above THE CIRCINUS GALAXY: A NEARBY GALAXY WITH AN EXPLODING HEART

At a distance of 13 million light-years, the Circinus Galaxy is one of the closest active galaxies to us. It is a Seyfert galaxy with an unstable, energetic nucleus, harboring a black hole. Rings of gas are being ejected from the galaxy's center, and measure some 1,400 light-years across. They have been cast out by the central black hole.

Above **M77: PORTRAIT OF A SEYFERT GALAXY**
This composite image shows reddish X-ray data
taken with the Chandra X-ray Observatory and
visible light imagery made with the Hubble Space
Telescope. Radio emission is shown in blue. The
galaxy is M77, one of the brightest Seyfert galaxies
in the sky, which lies 47 million light-years away
in Cetus. The X-ray emission comes from the black
hole within the galaxy's active nucleus.

Overleaf **THE SAUCER-SHAPED DISK OF THE
SOMBRERO GALAXY**
A popular target for backyard stargazers, the
Sombrero Galaxy (M104) in Virgo is adorned
by a nearly edge-on disk that appears like
the archetypal UFO. The prominent dust lane
running along the galaxy's edge is visible in
telescopes; the whole assemblage lies some
29 million light-years off.

Opposite NGC 2276: A DISRUPTED SPIRAL GALAXY WITH A BLACK HOLE
The spiral NGC 2276 (left) lies at a distance of 120 million light-years in the constellation Cepheus, and is adjacent on the sky to the elliptical galaxy NGC 2300 (right). NGC 2276 has an enhanced rate of star formation and contains an intermediate black hole, with a mass of 50,000 Suns, in one of its arms.

Above **VISITING CYGNUS X-1, THE FIRST BLACK HOLE**
In the early 1970s, Cygnus X-1 emerged as the first strong candidate for a black hole. A stellar black hole located within the Milky Way, its confirmation by 1990 set astronomers off on a path of discovery for black holes in other galaxies too.

SEYFERT GALAXY M77 AND ITS ERUPTIVE NUCLEUS
One of the brightest Seyfert galaxies in the sky,
M77 lies some 47 million light-years away in the
constellation Cetus. Its bright, active nucleus is easily
visible in backyard telescopes.

Opposite THE BLACK HOLE IN THE CENTER OF THE MILKY WAY
Using the Chandra X-ray Observatory, researchers have imaged the region surrounding the central black hole in our galaxy and have found it may be emitting neutrinos, particles with nearly zero mass and with no electoral charge. This image shows the region around Sagittarius A*, the energetic region surrounding the 4-million-solar-mass black hole. Bluish and orange plumes in the regions are the remnants of gas belched out by the area of the black hole millions of years ago.

Above X-RAY HOT SPOTS NEAR THE GALACTIC CENTER
An astonishing image made with the Chandra X-ray Observatory shows the area of the Milky Way's center in X-ray emission. This sprawling montage covers an area of about 900 by 400 light-years, and depicts high-energy objects like neutron stars, white dwarfs, and black hole emission, all bathed in an incandescent fog of gas glowing at millions of degrees.

Overleaf STARS AT THE GALAXY'S CENTER
The center of the Milky Way lies hidden behind thick veils of dust and gas. We can normally see stars and nebulae about $1/3$ of the way toward the Milky Way's core. But infrared light allows seeing the tangle of stars and gas in the crowded galactic center, some 26,000 light-years away. This picture's width is about 900 light-years, and it is filled with hot, young stars and clouds of reddish hydrogen gas.

Chapter Four

THE VIRGO SUPERCLUSTER

✳ ✳ ✳

Gravity's powerful force plays a significant role in the universe we see. But the percentage of galaxies that are interacting with each other is tiny—space, as the old saying goes, is *really big*. Most of the space between galaxies, in fact, is empty, with none of the glowing matter that we see in galaxies. As we move farther out in the cosmos, away from our Local Group of galaxies, we see that gravity plays an important role in the lives of all galaxies. Most galaxies exist, in fact, in groups like the Local Group and as members of clusters containing hundreds or thousands of galaxies. The Milky Way is no exception.

Let's journey on that spaceship of the imagination beyond the Local Group of galaxies. We have already seen that the diameter of our galaxy's disk is 100,000 light-years. We know that our Local Group of galaxies stretches

FAST FACT
The Virgo Cluster forms the heart of the far larger Virgo Supercluster of galaxies, which contains at least 100 galaxy clusters and groups, including our own Local Group.

some 10 million light-years across, and that traversing it end to end at the speed of light, in our spaceship, would take 10 million years. But let's now travel farther out to the biggest cluster of galaxies in our part of the universe, the Virgo Cluster. This group of galaxies is centered more than 50 million light-years away. If we could travel in our spaceship at the speed of light, it would take 50 million years to get there. The light we see from these galaxies today, in our telescopes, has been traveling through space since long before our human ancestors existed, since life on Earth had only begun to include the first bats and many groups of modern mammals, such as tapirs, rhinoceroses, and camels.

Back on Earth, in our nighttime sky, veteran observers know that the spring-time evening sky offers a particularly good time to see lots of galaxies. Dozens of galaxies spanning the whole array of types outlined by Hubble and de Vaucouleurs are splayed across the spring sky, concentrated in the constellations Virgo, Coma Bere-nices, Leo, and Canes Venatici and surrounding areas. By looking in this direction of the sky, we are looking away from the plane of our own galaxy, which obscures the distant galaxies beyond, and through a window onto a more distant cosmos. And it so happens that many bright galaxies—including M60, M61, M94, M96, M104, NGC 4535, and NGC 4762—are concentrated in this part of the sky because they belong to what astronomers term the Virgo Cluster, the largest group of galaxies in our little part of the universe.

FAST FACT

Discovery of the first galaxies in the Virgo Cluster dates to the 1770s, when the French comet observer Charles Messier found many of the brighter members, though he believed that they were nebulae.

FINDING OUR WAY INTO VIRGO

Our understanding of the largest groups of galaxies around us began to get traction in the middle of the twentieth century. As astronomers mapped more and more galaxies across the sky and deciphered their distances using redshifts, they discovered that a cloud of galaxies lay concentrated toward the constellation Virgo. This so-called Virgo Cluster contains at least 1,500 galaxies, and its center floats approximately 54 million light-years away, with a core populated by bright, supermassive elliptical galaxies, including M84, M86, M87, M49, and others. Many of the galaxies in the cluster are so bright that they display wonderful details even in the eyepieces of telescopes used by amateur astronomers, if viewed under a dark, moonless sky. The ellipticals are balanced by an equal number of spiral and barred spiral galaxies, and weird irregular galaxies also can be found, including some interacting pairs. For backyard astronomers, the Virgo Cluster offers one of the greatest telescopic playgrounds in the sky.

THE MEMBERS OF THE VIRGO CLUSTER

The Virgo Cluster is some five times more distant than the diameter of the Local Group. Its mixture of galaxies is pretty random, and the main mass of the cluster is oblong, oriented along an axis facing us on one end and facing away from us on the other, with its long axis four times as wide as its short side. The spirals are arrayed along this tube-shaped corridor, and the ellipticals are more generally concentrated toward the cluster's center. The cluster has three gravitationally dense "clumps": one centered on the supermassive galaxy M87; another concentrated around another elliptical, M86; and a third centered on the elliptical galaxy M60. The largest clump is the one centered on M87: it contains at least 100 trillion solar masses of material, making it the massive core of the cluster.

FAST FACT
Elliptical galaxy M86 is one of the central galaxies in the Virgo Cluster. It has the highest "blueshift," meaning the galaxy is moving toward us as opposed to away from us ("redshift"), of any Messier object, approaching us at a velocity of 244 kilometers per second.

KING OF THE HILL: BIG BAD M87

The largest individual galaxies in the Virgo Cluster are amazingly diverse. The single most dominant galaxy in the cluster is M87, one of the largest elliptical galaxies known. It is classed as one of the rare so-called cD (centrally dominant) galaxies—supergiant ellipticals that are the big guys in rich galaxy clusters. Discovered by French astronomer Charles Messier in 1781, the galaxy has been known as a giant ever since twentieth-century astronomers have been studying it. The halo of stars and gas making up M87's spherical form stretches nearly 500,000 light-years across, dwarfing the Milky Way Galaxy.

Although it is a relatively formless giant ball of stars, M87 is notable for showing a jet protruding from one side of the galaxy that is so bright it's visible even in some images taken by amateur astronomers. This is the jet of material being slingshot from the matter that is falling toward the supermassive black hole in the galaxy's center but avoiding falling into it. The high-energy "screams" of this material being shot outward at extremely high velocities are visible in X-rays and gamma rays. The jet twists along its inner path as material moves away from the black hole's accretion disk and is focused into a beam that extends to the incredible distance of 250,000 light-years.

The supermassive black hole that powers M87's jet is one of the largest known. It contains an estimated mass of 5 to 7 billion suns, as opposed to the 4.3-million-solar-mass black hole in our Milky Way's core. M87's black hole, then, is more than 1,000 times more massive than ours in the Milky Way. In 2019 astronomers released an image of the region of this black hole, the first made with the Event Horizon Telescope. M87 is also widely known for its huge population of globular star clusters that orbit far out in the galaxy's halo. Astronomers believe that M87 holds some 12,000 of these clusters, compared with the Milky Way's population of around 150.

FAST FACT

M87, one of the most massive galaxies known, has a population of 12,000 globular star clusters, as well as a central black hole "weighing" 7 billion solar masses.

The supermassive black hole that powers M87's jet is one of the largest known.

THE VIRGO SUPERCLUSTER

NGC 5005 group
Distance: 60 million ly

NGC 4565 group
Distance: 57 million ly

Virgo cluster
Distance: 55 million ly

NGC 5746 group
Distance: 64 million ly

Ursa Major North group
Distance: 60 million ly

NGC 4274 group
Distance: 53 million ly

M61 group
Distance: 57 million ly

NGC 3665 group
Distance: 50 million ly

NGC 5364 group
Distance: 50 million ly

NGC 4179 group
Distance: 52 million ly

Ursa Major South group
Distance: 56 million ly

NGC 5775 group
Distance: 52 million ly

NGC 4666 group
Distance: 51 million ly

NGC 5084 group
Distance: 60 million ly

NGC 5866 group
Distance: 50 million ly

Canes Venatici II group
Distance: 30 million ly

NGC 4697 group
Distance: 40 million ly

NGC 3585 group
Distance: 64 million ly

M66 group
Distance: 33 million ly

M96 group
Distance: 36 million ly

M104 group
Distance: 36 million ly

M51 group
Distance: 26 million ly

Canes Venatici I group
Distance: 15 million ly

NGC 5121 group
Distance: 52 million ly

NGC 2273 group
Distance: 58 million ly

M101 group
Distance: 18 million ly

NGC 3175 group
Distance: 48 million ly

M81 group
Distance: 12 million ly

Centaurus A group
Distance: 12 million ly

0°

NGC 2997 group
Distance: 37 million ly

NGC 4976 group
Distance: 47 million ly

90°

IC 342 group
Distance: 11 million ly

NGC 2835 group
Distance: 33 million ly

270°

20 million ly

LOCAL GROUP

40 million ly

180°

NGC 7331 group
Distance: 42 million ly

NGC 1023 group
Distance: 32 million ly

Sculptor group
Distance: 12 million ly

IC 5181 group
Distance: 47 million ly

NGC 1672 group
Distance: 39 million ly

IC 5332 group
Distance: 32 million ly

NGC 1808 group
Distance: 42 million ly

M77 group
Distance: 40 million ly

NGC 7582 group
Distance: 61 million ly

NGC 1433 group
Distance: 55 million ly

Dorado cluster
Distance: 59 million ly

NGC 1084 group
Distance: 56 million ly

NGC 1097 group
Distance: 46 million ly

Fornax cluster
Distance: 62 million ly

NGC 908 group
Distance: 62 million ly

NGC 681
Distance: 63 million ly

NGC 1255 group
Distance: 65 million ly

THE HEART OF
THE VIRGO CLUSTER

Near M87 in the sky lies a line of bright galaxies that inhabit the inner part of the Virgo Cluster, which includes M84 and M86, bright galaxies discovered in 1781 by Messier. This string of galaxies also includes NGC 4477, NGC 4473, NGC 4461, NGC 4458, NGC 4438, and NGC 4435. This line of galaxies is Markarian's Chain, which marks the observational core of the cluster. The Armenian astronomer Benjamin Markarian discovered the common motion of these galaxies in space in the early 1960s. M84 is an elliptical galaxy with a prominent pair of dust lanes that can be seen crossing the face of the galaxy. It lies at a distance of 60 million light-years. Nearby, M86 is another massive elliptical galaxy with an intensely bright center, and it lies some 52 million light-years away. Both galaxies and their neighbors in Markarian's Chain make up the core of the Virgo Cluster galaxies, on which many amateur observers have concentrated their efforts. Other unusual nearby galaxies include M60, M61, M85, M89, M90, M91, M100, and NGC 5033.

FAST FACT
The Virgo Cluster of galaxies is the nearest large cluster of galaxies to us, comprising at least 1,300 galaxies in a sphere measuring at least 15 million light-years across.

HERE COME THE
SUPERCLUSTERS

Opposite **THE LOCAL SUPERCLUSTER OUT TO 65 MILLION LIGHT-YEARS** The Local Group is but one of many galaxy collections belonging to the Virgo Supercluster. This vast assemblage of relatively small groups and large clusters extends more than 50 million light-years from its center in the massive Virgo Cluster. In this illustration, we show the part of the supercluster that runs from the Local Group (located at the map's center but in reality near the supercluster's edge) to slightly beyond the Virgo Cluster. This graphic contains all galaxy groups and clusters containing at least three relatively large galaxies.

The Virgo Cluster that skygazers are familiar with is just part of the story of the local universe near us. Galaxies exist in groups (as with our Local Group), in clusters (as with the Virgo Cluster), and also in much larger aggregations. So-called superclusters hold many thousands of galaxies and exist on a scale an order of magnitude larger yet. The Virgo Supercluster, also called the Local Supercluster, is the largest group of galaxies to which our Milky Way and most of the galaxies we see easily in our sky belong. It is monstrously larger than the Virgo Cluster itself.

The Virgo Supercluster contains some 100 galaxy groups and clusters altogether and has a diameter of approximately 110 million light-years. Some 10 million such superclusters make up all of the galaxies that astronomers can see in the entire visible universe.

Understanding the Virgo Supercluster really begins with numerous observations made by the great German-English astronomers William Herschel and his son John Herschel. Their records of numerous nebulae in the area of Virgo, Coma Berenices, and surrounding constellations suggested a major concentration and set astronomers to wondering about the great many fuzzy nebulae that appeared in this region of sky. By the time John Herschel published his *Catalogue of Nebulae and Clusters of Stars* in 1863, the world had a pretty good survey of the numbers of "spiral nebulae" and other strange nebulae littering that region of the sky, most of which turned out to be galaxies.

FAST FACT
The three major subgroups of the Virgo Cluster will eventually merge, forming a tighter group of interacting galaxies, drawn together by gravity.

VIRGO DRAWS IN ASTRONOMERS

It wasn't until a generation after Hubble's breakthrough discovery that astronomers began to realize the significance of the concentration of galaxies toward Virgo. In the early 1950s, Gérard de Vaucouleurs published several papers in which he argued that the excess of galaxies in this region represented a large-scale galactic structure of some sort. In 1953, de Vaucouleurs created the term "Local Supergalaxy" to explain the concentration, and five years later he coined the term "Local Supercluster." Not to be outdone, Harlow Shapley suggested the term "Metagalaxy," but eventually Local Supercluster, and then Virgo Supercluster, won the day, even as astronomers argued about whether the concentration of galaxies was significant or merely a chance alignment.

By the 1970s, astronomers had made enough progress with large-scale surveys of redshifts to understand that the concentration of galaxies in the direction of Virgo was real—most of the objects lay at similar distances, and an enormous cloud of galaxies did indeed exist in this direction of the sky. They studied many of these galaxies, ones like NGC 7331 and NGC 7814.

AN ELLIPTICAL FOOTBALL

The next leap forward in understanding the Virgo Supercluster came in a landmark paper in 1982 by the Canadian-American astronomer R. Brent Tully. He published an extensive analysis of the Virgo Supercluster, suggesting that it contains a flattened disk of approximately two-thirds of the supercluster galaxies, in addition to a spherical halo holding the remaining third of the galaxies. In this sense, the basic structure of the supercluster is somewhat like the structure of a spiral galaxy itself, in the simplest terms, and is an ellipsoid—roughly football-shaped. Tully suggested that the disk component itself is quite thin, perhaps only 3 to 5 million light-years "high," and with a long dimension that is at least six times greater than its short dimension. Numerous galaxies served as examples in these studies, such as NGC 253, NGC 2903, and IC 356.

In the first years of the twenty-first century, Australian astronomers released a set of data from a project called the 2dF Galaxy Redshift Survey, conducted with the 3.9-meter telescope at the Australian Astronomical Observatory. Made public in 2003, this information offered a view of the large-scale universe in two "slices" out to about 2.5 billion light-years. The enormous amount of data allowed astronomers to compare the Virgo Supercluster with several other nearby superclusters for the first time. They found that the Virgo Supercluster is rather "poor," meaning that it lacks a concentrated center. Our supercluster is also quite small compared with many others observed at greater distances. The single "rich" galaxy cluster, the Virgo Cluster, lies near its center, and the filaments of galaxies and galaxy groups surrounding it are rather sparse compared with many richer clusters in the universe.

The Virgo Supercluster is poor, lacking a concentrated center.

HOW WE FIT
INTO THE COSMOS

Just as the Sun and our solar system lie far out from the Milky Way's center, in the galactic "suburbs," our Local Group lies pretty far out, significantly away from the central action of our own supercluster. The Local Group lies along a small filament of galaxies that extends from the Fornax Cluster of galaxies to the Virgo Cluster. And the immense scale of the Virgo Supercluster adds to our sense of being isolated, in a small filament out to one side: overall, the supercluster has a volume some 7,000 times greater than the Local Group, and 100 billion times greater than our galaxy alone. Again, space is *really big*!

WHERE ARE WE?
Astronomers use different coordinate systems to give object positions. In the solar system, Earth's orbital plane (the ecliptic) is a good reference. Beyond that, the Milky Way's plane is best. The two planes meet at an angle of 60°.

PLANE OF THE MILKY WAY

90°

180°

SUN

ECLIPTIC PLANE

0°   60°

270°

With a telescope, look at how galaxies are spread around the sky. You might target galaxies like NGC 1365, NGC 7217, and NGC 7479. The distribution of galaxies within the Virgo Supercluster is not uniform. The vast majority of bright galaxies lie relatively close to the supercluster's center and fall off dramatically as distance from the center increases. The greater number of bright galaxies lie within a small number of clouds within the supercluster. Nearly all of the galaxies are found within the Canes Venatici Cloud, the Virgo Cluster, and the other nine significant clouds: Virgo II, Leo II, Virgo III, Crater, Leo I, Leo Minor, Draco, Antlia, and NGC 5643.

THE BIG NOTHING

What this means is that the vast majority of the volume in the Local Supercluster exists as a great void, an area lacking galaxies. And the distances between our Virgo Supercluster and other superclusters of galaxies are vast, separated by immense voids. The diameters of these voids range from dozens to hundreds of millions of light-years, with filamentary chains of galaxies twisting their way around the voids. On very large scales, galaxies in clusters and superclusters might be thought of as soap bubbles, with galaxies coating the surfaces and vast spaces of nothingness lying in between. To appreciate the different forms of galaxies, check out examples like Copeland's Septet, Fornax A, M65, M66, M77, M88, M90, M95, M108, NGC 660, NGC 772, NGC 1055, NGC 1097, NGC 1313, NGC 1672, NGC 3717, NGC 4151, NGC 4365, NGC 4725, NGC 5907, NGC 7331, IC 1613, and IC 2574.

> The vast majority of the Local Supercluster exists as a great void.

Before we look at the very large-scale universe, however, consider some of our neighbors lying just beyond the Local Group, and closer than the center of the Virgo Cluster itself. These galaxies are of great interest to astronomy enthusiasts because they constitute many of the galaxies that can be viewed in backyard telescopes and can also be captured with cameras. Many of these relatively close galaxies give us a spectacular picture of the galaxy types that must exist throughout the entire cosmos.

NEARBY GALAXY GROUPS

Moving out from our own Local Group, we can encounter several other small groups of galaxies relatively nearby. The closest is the IC 342/Maffei 1 Group, which contains at least eighteen galaxies. Several of the galaxies in this binary group are concentrated around the face-on barred spiral IC 342, and the other subgroup is concentrated around the elliptical galaxy Maffei 1, some 9 million light-years away. Maffei 1 and a nearby neighbor, Maffei 2, lie in a rich area of the Milky Way in the constellation Cassiopeia, as seen in our sky; they are so faint and obscured by dust in our own galaxy that they were undetected until the Italian astronomer Paolo Maffei recorded them in 1967.

Another close galaxy group, the M81 Group, lies some 11 million light-years away and contains at least thirty-four galaxies. The brightest two, M81 and M82, are familiar to backyard skygazers, as they are bright galaxies in the springtime sky, lying in the constellation Ursa Major. M81, sometimes called Bode's Galaxy because it was discovered by the German astronomer Johann Bode, in 1774, is a brilliant, highly inclined spiral galaxy visible even through a pair of binoculars. A short distance away from M81, visible in the same low-power telescopic field of view, lies the distorted irregular galaxy that is visible nearly edge-on to our line of sight. Sometimes called the Cigar Galaxy, M82 is an amazing object because it is undergoing an eruptive starburst episode, with energetic bouts of star formation, perhaps triggered by gravitational interaction with M81. The M81 Group also contains the bright and notable galaxies NGC 2403, NGC 2976, and NGC 3077.

FAST FACT
The heart of the Virgo Cluster, a curved line of galaxies that includes M84, M86, and six other bright galaxies, is called Markarian's Chain.

As astronomers continue to study what makes up the Virgo Supercluster, they will eagerly study groups of galaxies within it that lie at increasing distances. Continuing outward from our home base, the next galaxy group we encounter is the Centaurus A/M83 Group, which contains some really unusual objects. This group lies at a distance of between 12 and 15 million light-years and contains two subgroups, one centered on Centaurus A and the other on M83. This group contains twenty-nine members gravitationally wandering around Centaurus A and another fifteen in the vicinity of M83.

Centaurus A may foreshadow what the Milky Way will look like when it merges with the Andromeda Galaxy.

———•———

Centaurus A is a bright galaxy located in the southern sky that can be seen easily in amateur telescopes; it has a peculiar designation because it was first discovered to be a strong radio source. The radio emission from this peculiar, jumbled galaxy emanates from the galaxy's central black hole, which "weighs" 55 million solar masses. The galaxy is the result of a major collision and merger between two galaxies in the distant past, and this huge, energy-laden elliptical cloud may presage the future of the Milky Way, when our galaxy merges with the Andromeda Galaxy. Jets blasting away from the galaxy's center, carrying material that escaped falling into the black hole, can be imaged by backyard astronomers with a proper camera setup.

The other lobe of this galaxy group is the one centered on the bright galaxy M83, which is another familiar sight to astronomy enthusiasts. This face-on barred spiral galaxy lies in the constellation Hydra, also in the southern sky, and approximates what the Milky Way generally looks like, if we could see our galaxy from the outside. The M83 galaxy's disk is only some 60,000 light-years across, so it might be a two-thirds scale model of the Milky Way.

OUT INTO DEEPER SPACE

The next galaxy group, moving outward, is the Sculptor Group, which lies about 13 million light-years away and also is positioned in our southern sky from the viewpoint of Earth. This group of at least thirteen galaxies contains one of the brightest edge-on spirals, NGC 253, sometimes called the Sculptor Galaxy. This beautiful object is a starburst galaxy, like M82, undergoing a powerful burst of star formation. Its central black hole consists of about 5 million solar masses, making it slightly beefier than the one in our Milky Way. Also in this group are the bright galaxies NGC 247 and NGC 7793; the well-known galaxies lying near these in the sky, NGC 55 and NGC 300, are thought to be foreground galaxies lying at closer distances.

Slightly farther away than the Sculptor Group, and aiming back into the northern sky as we see it, is the Canes Venatici I Group, also called the M94 Group. Lying some 13 million light-years away, this group contains at least fourteen members centered on the multi-ringed spiral galaxy M94. This galaxy has an intensely bright nucleus and is well known to amateur observers; it has an inner ring and a distinct outer ring that is a complex arrangement of spiral arms.

And then comes the NGC 1023 Group, a small group of at least five galaxies at a distance of 21 million light-years. This diminutive group contains some objects that are well known to observers, not only the lenticular galaxy NGC 1023 itself but the edge-on spiral galaxy NGC 891 and the barred spiral NGC 925.

And the galaxy groups just keep on coming. Next in line outward is the M101 Group, also 21 million light-years away, a small group of at least seven galaxies centered on the big, bright face-on spiral galaxy M101. This galaxy is also well known to amateur observers, being one of the brightest galaxies in Ursa Major, located near the Big Dipper asterism in the sky. M101 is often called the Pinwheel Galaxy, owing to its pinwheel formation of face-on spiral arms. This is a physically large galaxy: its bright disk stretching 170,000 light-years across makes it some 70 percent greater than the Milky Way's. Its arms appear asymmetrical, and a large number of star-forming regions are scattered along the galaxy's spiral arms.

HALFWAY TO THE VIRGO CLUSTER

Located some 25 million light-years away is the NGC 2997 Group, a small collection of galaxies swarming the Southern Hemisphere galaxy NGC 2997 in Antlia. More groups lie a short distance farther away: the Canes Venatici II Group contains the bright galaxy M106 and a variety of smaller galaxies. Similarly, the M51 Group contains a collection of galaxies in the vicinity of the Whirlpool Galaxy, M51, one of the bright showpiece galaxies known by amateur observers. At this distance, 31 million light-years, the catalog of galaxy groups has stretched halfway to the center of the Virgo Cluster itself. Exploring these galaxies gives you a pretty good feeling about the kinds of galaxies that lie close by in our cosmic neighborhood.

But how does the nature of galaxies change when we venture much farther away from home? What are galaxies like as we move far away, out to the edges of space and time?

M101 is often called the Pinwheel Galaxy, owing to its face-on spiral arms.

NEW IDEAS
ABOUT GALAXY MERGERS

In recent years, astronomers have continued to study colliding galaxies well enough that they have created a collision theory, which defines certain types of collisions and what is likely to have produced the strange forms we now see. The first and most dramatic type is the so-called head-on collision, akin to two trucks slamming into each other on a freeway. This catastrophic event produces such dramatic objects as the Cartwheel Galaxy, a "ring galaxy" lying 500 million light-years away in the constellation Sculptor that appears to consist of a strongly condensed nucleus surrounded by a bright ring of star formation, with little material between the center and the ring. As galactic blows might be classed, this was a direct, head-on smash to the face.

Fritz Zwicky discovered the Cartwheel Galaxy in 1941, and over the ensuing decades astronomers found many more ring galaxies, although they are comparatively rare. In the 1970s, astronomers using computer modeling explained the basic physics of ring galaxies. They result when an intruder galaxy smacks right into a target galaxy—one with a large disk containing stars and gas clouds in circular orbits—along its axis of rotation, falls straight through the disk, and comes out the back end. As collisions go, it's a bull's-eye.

THE RANGE OF RING GALAXIES

The Cartwheel Galaxy resulted from a pretty symmetrical, head-on blow. But of course perfect symmetry is rare in the universe, and so a whole range of ringlike galaxies can be produced from galaxies that strike offset from the exact center, and without purely perpendicular velocity. So a numerous range of asymmetrical ring galaxies are left with squished rings, blobby centers, warped arms, and all manner of other oddities.

One particularly amazing example of a ring galaxy is Hoag's Object, named for the American astronomer Arthur Hoag, who discovered the galaxy in 1950. Lying some 600 million light-years away in the constellation Serpens, this object consists of a nearly perfect ring of young, hot blue stars and gas surrounding an intensely yellowish, compact nucleus. Hoag's Object may be the result of a collision, but no one yet knows for sure. In a perfect collision, the ring forms from a density wave in the galaxy's disk that leads to the ringlike appearance. But in the case of Hoag's Object, there is no sign of the impactor—the speeding bullet, if the galaxy is a merger, seems to have left the scene.

Other types of ring galaxies do exist, the most intriguing of which are the polar ring galaxies. A good example is NGC 4650A, which shows a ring of material at right angles to a bright, lenticular galactic center. This type of galaxy results from a collision that pulls much of the galaxy's gas out into a new configuration, leading to this odd appearance.

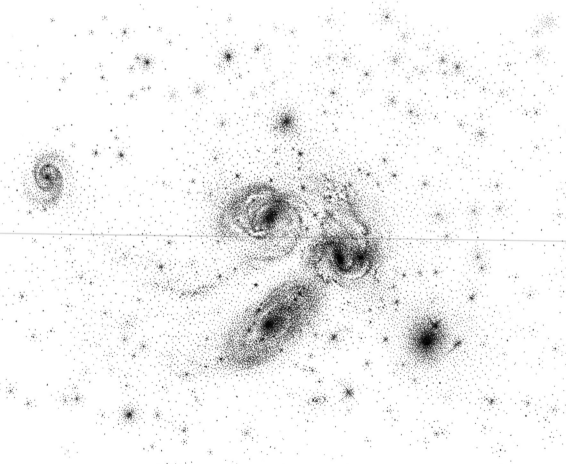

HOW GALAXIES MERGE OVER TIME

Although the concept of trekking outside to view and study interacting galaxies is a relatively new one, the realization that galaxies sometimes collided goes back quite some time. Soon after Hubble's discovery of the nature of galaxies, he and several other astronomers began to ponder whether these huge star systems interacted. Among the pioneers in the 1920s to think about such things were Harlow Shapley and the Swedish astronomer Bertil Lindblad. Right away, these astronomers and others concentrated on the likeliest places to find galaxies interacting with each other—the groups and clusters. One of the most amazing examples of interaction that backyard astronomers know about in their telescopes is Stephan's Quintet, a group of several galaxies in Pegasus.

As astronomers have begun to study greater numbers of galaxies, it has become clear that only a small percentage of them are interacting. But these systems, showing dynamical collisions, offer much for astronomers to learn. Roughly half of all galaxies exist in clusters, and the dark matter halos surrounding most galaxies extend far beyond the visible disks. So more interaction is taking place than we might think. Galaxy interactions also take a long time to occur, from hundreds of millions to a few billion years. So in order to see galaxy interactions, we need to catch them at the right time: during the 1 to 10 percent of the galaxy's life when the encounter takes place. Rich clusters of galaxies like the Coma Cluster give astronomers a good laboratory for seeing such interactions.

FAST FACT

The small interloping galaxy NGC 5195 is speeding past the Whirlpool Galaxy, M51, and drawing material away from it. It is a great example of an interacting galaxy visible in backyard telescopes.

TIDAL INTERACTIONS
BETWEEN GALAXIES

Unlike a direct collision that produces a ring galaxy, wherein the direct pull of the companion's gravity rearranges the result, most galaxy interactions happen in less precisely targeted ways. A large number of galaxies, particularly those in groups and clusters, interact from tidal effects that change one or more spiral arms or distort just one area of a galaxy as one passes the other in space.

The Whirlpool Galaxy is a good example of a tidal interaction between galaxies. The smaller galaxy, NGC 5195, is speeding by the larger galaxy, M51, and this scene is bright enough to be viewed in backyard telescopes. In the 1970s, the Estonian-American astronomer Alar Toomre and his brother Jüri Toomre, an astrophysicist, produced computer simulations that helped them explain many types of galaxy interactions involving tidal tails. They identified the processes that led to the tidal tails visible in M51 and also in galaxies like the Antennae and the Mice.

Similar galaxies were scrutinized and explained in later years. In the 1990s, Debra and Bruce Elmegreen studied the interacting pair NGC 2207 and IC 2163, sometimes called "the Eyes." The Elmegreens, along with the American astronomer Curtis Struck and others, studied this pair and found that it represents a specific class with a "pinched eyelid" appearance and short tidal tails that is often called an "ocular galaxy." The ocular appearance results from a phase wave that propagates through the galaxy in so-called fly-by collisions. Examples like the Antennae galaxies and the NGC 3190 Group provide interesting looks at galaxy interactions.

TAILS, BARS, AND
GRAVITATIONAL DISRUPTIONS

Other types of tidal interactions take place too. Some can create bars in galaxies. Depending on geometry and velocities, some tails can extend for very long distances, swept out by high-speed dwarfs. Some tidal interactions can distort galaxies so much that they are left in the shape of a guitar (Arp 105), or they can produce retrograde encounters—wherein the bullet galaxy flies by in the opposite direction of the main galaxy's disk rotation.

One prominent galaxy that has multiple arms and has been distorted in retrograde fashion is the Pinwheel Galaxy, M101 in Ursa Major, with its lopsided disk. In this case, the interloping galaxy is not known, but M101 certainly seems to have been affected by an encounter. It's possible that the slightly distorted nearby galaxy NGC 5474 danced with the bigger galaxy. Other galaxies without obvious "current" interaction also show distorted effects, as with the many-armed spiral NGC 4622, in which one set of spiral arms rotates in the opposite direction relative to other arms.

Tidal interactions occur when galaxies encounter each other and draw material outward, distorting galactic shapes. These are visible in many colliding galaxies, such as in the NGC 68 Group, the NGC 708 Group, and Seyfert's Sextet. But what happens when galaxy interactions are more catastrophic? Head-on collisions create mergers that often totally remake both merging galaxies. Some astronomers believe that elliptical galaxies form from the mergers of spiral galaxies, and the Toomres are big proponents of this idea. After all, most elliptical galaxies lie in clusters where galaxies are packed in and there have been lots of opportunities for mergers. The Toomres and other astronomers have studied what came to be known as the "merger rate"—a way of understanding the evolution of galaxies that have merged in clusters and other environments. More interesting examples lie in Arp 147, Arp 272, Arp 273, ESO 510-G13, NGC 5216, and NGC 6745.

GALAXIES BUILD UP FROM MERGERS

The significance of mergers in the creation of elliptical galaxies will continue to be studied for a long time to come. But it is already clear that galaxies build themselves via mergers, and our own galaxy is no exception to the rule. It's probable that the Milky Way consists of the remains of as many as 100 small galaxies that came together over the 9 billion years the galaxy has existed.

Taking this to the extreme, consider the Hubble Deep Fields. The Hubble Space Telescope has taken a series of "deep fields"—extremely long exposures of small pieces of sky, made in order to image the faintest galaxies possible. The first deep field was made in 1995, concentrating on a

Galaxies build themselves via mergers, and our galaxy is no exception to the rule.

small area of sky in Ursa Major. Nearly all of the 3,000 objects in the field are galaxies. Later, in 2004, astronomers released the Hubble Ultra-Deep Field, an even longer exposure that shows about 10,000 galaxies in a small area of the constellation Fornax. And later still, in 2012, they released the Hubble eXtreme Deep Field—a refined version of the Ultra-Deep Field that shows galaxies that formed just half a billion years after the Big Bang.

The bottom line from these images is that they show numerous small, blobby galaxies—the protogalaxies that came together, as time marched on, to form normal galaxies like ours that we see close by in the modern universe. Remember that as we look farther away in the cosmos, into images like the deep fields, we're also looking back in cosmic time.

REMAKING GALAXIES
WITH STARBURSTS

In the last decade or two, astronomers have learned a great deal about the details of galaxy collisions. When galaxies merge, energy gets imparted into their disks, and they often display renewed star formation. If the merger happens in an energetic way, astronomers call this a "starburst event." You can see bursts of star formation happening in many merging galaxies. Gas is compressed and gravity takes over, generating new clusters of stars, sometimes on a wide scale. A good example is the Cigar Galaxy (M82), in Ursa Major. This bright galaxy, which is easy to see in backyard telescopes, is undergoing an energetic starburst event right now. The bright nearby galaxy M81 is gravitationally tugging on M82, causing the starburst event. More unusual cases can be found in Arp 302, NGC 3169, and NGC 4676A and B.

Mergers can also have catastrophic effects on the centers of galaxies. Astronomers now know that nearly every galaxy, except for dwarfs, has a central supermassive black hole. As we have seen, our own Milky Way harbors one. When young, these black holes were active engines, sucking down all of the matter nearby and flinging outward the energy and matter that didn't fall straight in, creating a brilliant quasar nucleus. Then, as they ran out of material to ingest, the black holes became quiescent and went to sleep.

A merger can dump vast amounts of stars and gas into formerly sleepy black holes, reawakening them into a fury.

WAKE UP, BLACK HOLES!

A merger can dump vast amounts of stars and gas into formerly sleepy black holes, reawakening them into a fury. This might restart or create a quasar nucleus in the center of a galaxy, forming an energetic galactic monster. This may be happening in the nearby galaxy merger Centaurus A (NGC 5128), a beautiful galaxy in the southern sky that is visible in small telescopes. The two black holes in a galaxy merger will swirl together and merge, creating a more massive central black hole that will reset the dynamics of the galaxy's nucleus.

BLACK HOLES
AS CENTRAL GALAXY
ENGINES

JET

GAS CLOUDS

DUSTY TORUS

EVENT HORIZON

SINGULARITY

ACCRETION DISK

Of course, not all galaxy mergers are major, catastrophic events involving head-on crash scenarios. Vast numbers of galaxy interactions happen with two galaxies of greatly different sizes and masses. What happens when a large galaxy interacts with a small one? These so-called minor mergers are curious and show a wide range of effects. When the smaller galaxy has 10 percent or less of the larger galaxy's mass, the big one generally suffers relatively small effects. Check out the unusual examples in the Perseus Cluster of galaxies and Arp 227.

> Our own Milky Way will end up in a substantial merger several billion years from now.

GALAXY SHELLS

The smaller galaxy passing through the disk of a larger one imparts an effect called tidal shocking. This can create shells around galaxies, counter-rotating disks within galaxies, tidal tails, and other features. The well-known galaxy M64 in Coma Berenices, the Black Eye Galaxy, is a good example of a counter-rotating disk: the inner disk rotates in the opposite direction from the outer disk.

As we have seen, our own Milky Way Galaxy will end up in a substantial merger several billion years from now. When the Andromeda Galaxy and the Milky Way collide, our galaxy will become part of a local supergalaxy, "Milkomeda," and the night sky from planets within the galaxy will be lit up with new spectacular features that we can't see today. The variety of such galaxy interactions grows enormously as we move out deeper into the universe, as do the number of just plain old galaxies of all shapes and sizes.

Opposite **BLACK HOLES AS CENTRAL GALAXY ENGINES**
The majority of galaxies contains supermassive black holes with masses more than a million times that of the Sun in their centers. Sometimes these galaxies accrete matter rapidly toward their centers, throwing out lots of energy and attracting astronomers' attention. These are called active galactic nuclei, or AGN. When viewed from certain angles, AGN may look different from similar objects oriented a different way.

Black holes are spherical, but matter can fall in from any direction. Most infalling matter forms an accretion disk, which heats up and shines brightly. This is normally how astronomers detect a black hole. Black holes often produce jets—material pulled off the accretion disk and flung outward before it can fall past the event horizon.

A dusty "torus" is thought to surround the entire system. While it's often drawn as a dusty ring or doughnut, its actual shape is still in question. Additional clouds of material can also be caught in the black hole's gravity, moving faster or slower depending on their proximity to the central black hole.

**NGC 3949: A STRIKING COUSIN OF OUR
MILKY WAY GALAXY**
Because we cannot see our galaxy from the
outside, we must be content with viewing
similar galaxies in the nearby universe. One
such galactic cousin is NGC 3949, a barred
spiral about 50 million light-years away in
the constellation Ursa Major.

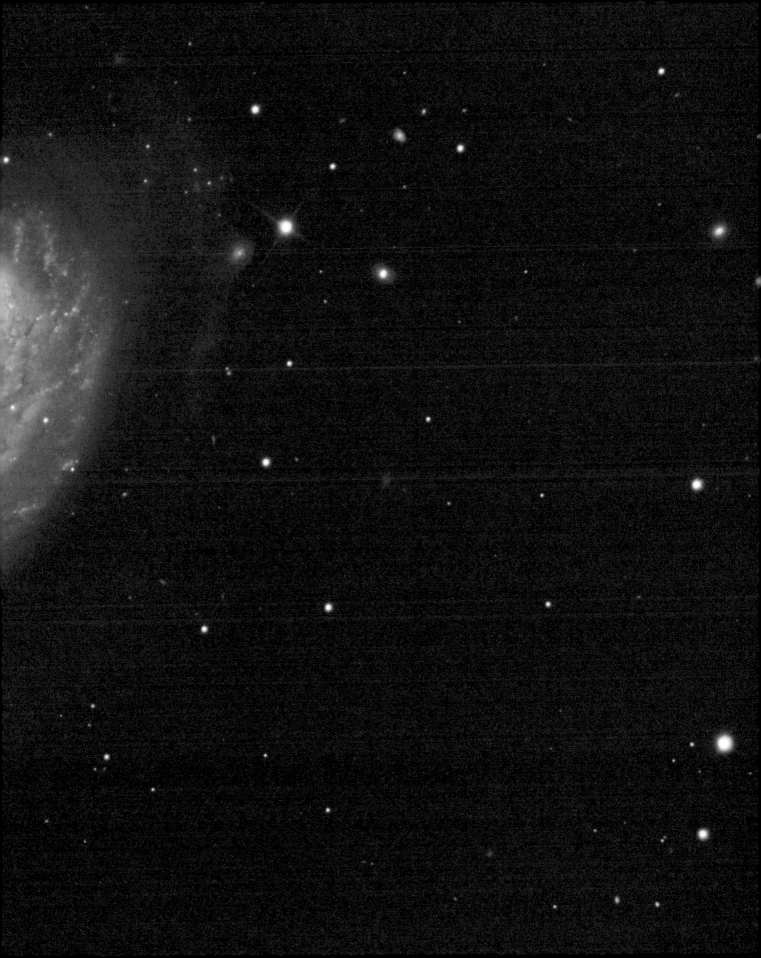

Previous **NGC 4725: A DUSTY, RINGED SEYFERT GALAXY**

The barred spiral galaxy NGC 4725 lies at a distance of 40 million light-years in the constellation Coma Berenices. It is a Seyfert galaxy with an energetic, active nucleus, and harbors a substantial supermassive black hole in its center.

Opposite **THE WHALE AND ITS LITTLE COMPANION**

The galaxy NGC 4631 in Canes Venatici, lying at a distance of 30 million light-years, has a distinctive shape that has given it the moniker Whale Galaxy. Its faint companion is NGC 4627, a dwarf elliptical galaxy that is circling the Whale. Eventually this little object will be gravitationally swallowed by the whale, causing a new round of star formation.

Above **M100 AND ITS FAMOUS 1979 SUPERNOVA**

Bright and easily visible in backyard telescopes, M100 in Coma Berenices is a favorite target for skywatchers. At 55 million light-years distance, it is a part of the Virgo Cluster of galaxies. In 1979 a bright supernova appeared in M100, visible in this image as the bright star just below the bluish knot of gas in the galaxy's lower spiral arm (near the bottom of the picture and left of center). Thirty years after the star exploded, the Chandra X-ray Observatory detected X-ray emission from the region, suggesting that the supernova has spawned the youngest black hole known in our region of the cosmos.

Overleaf **NGC 660: A CURIOUS POLAR RING GALAXY**

A rare type, polar ring galaxies have a large population of stars, gas, and dust in a ring, orbiting the main body of the galaxy, and oriented nearly perpendicular to the primary axis. NGC 660 in Pisces, lying at a distance of 45 million light-years, is such a galaxy. The polar ring appears alive with pinkish glows of star formation. The ring may consist of material captured from another galaxy that passed the scene long ago.

Opposite **THE HEART AND SOUL NEBULAE—AND TWO NEARBY GALAXIES**

Two large, sweeping gas clouds in Cassiopeia are nicknamed the Heart and Soul Nebulae. The Heart, IC 1805 (right), and Soul (IC 1848, left) lie in the same region as Maffei 1 and Maffei 2, which in this image are tiny glows below and between the two big nebulae.

Top **MAFFEI 1: A LOCAL GROUP IMPOSTOR**

For years astronomers believed the unusual galaxy Maffei 1 was a member of the Local Group. Now they believe it is just beyond the group's limits, at 9 million light-years away. This massive elliptical galaxy lies in the thick of the Cassiopeia Milky Way in our sky, meaning it is heavily obscured. If it were away from the Milky Way, it would shine far more brightly.

Bottom **MAFFEI 2: ANOTHER LOCAL GROUP IMPOSTOR**

Maffei 1 and 2 were discovered by the Italian astronomer Paolo Maffei in 1968, each heavily obscured by the intervening Milky Way stars, gas, and dust. The intermediate spiral Maffei 2 was thought for a time to be a Local Group member, just like its brethren, but now is known to lie just beyond it, at 9.8 million light-years' distance.

Previous **THE MARVELOUS FACE-ON SPIRAL IC 342**
At 11 million light-years, IC 342 is a close galaxy to ours, and is a member of a loose group along with several others. It has a low surface brightness, its average glow over individual parts, and lies in Camelopardalis, near the plane of the Milky Way in our sky. So it's a little difficult for amateur astronomers to spot, despite its elegant form.

Above **THE MAGNIFICENT PAIR OF GALAXIES M81 AND M82**
Two of the brightest galaxies in the springtime evening sky lie a short distance from each other, and are gravitationally interacting. They are M81, sometimes called Bode's Galaxy (left), and M82, sometimes called the Cigar Galaxy.

Opposite **THE BEAUTIFUL SPIRAL GALAXY M81 IN URSA MAJOR**
One of the brightest galaxies in the northern sky, M81 in Ursa Major, sometimes called Bode's Galaxy, lies 12-million light-years away and features a bright, active nucleus and shimmering arms, peppered by tiny pinkish regions of star formation. The galaxy hides a supermassive black hole with 70 solar masses in its center.

Opposite **THE OUTER SHELLS OF CENTAURUS A**

Centaurus A (NGC 5128) is a bright galaxy in the southern sky, lying in the sparkling constellation Centaurus. It was named for its nature as a strong radio source. Astronomers discovered that this big, ball-shaped galaxy is the result of an ancient merger of two galaxies that slammed together head-on. Thus, the object may hold a clue as to what "Milkomeda" will appear like, the future supergalaxy that will merge Andromeda with the Milky Way. Faint bluish shells around Centaurus A are indications of turbulence from the long-ago merger. The galaxy lies some 13 million light-years away.

Above **AN X-RAY LOOK INTO THE HEART OF M82**

M82 is a starburst galaxy with an eruptive and active black hole in its center. A deep image taken in X-rays with the Chandra X-ray Observatory shows the burst of star formation occurring near M82's center, with new stars winking on at rates hundreds of times greater than that of the Milky Way.

Above **THE SOUTHERN PINWHEEL GALAXY**
M83, a magnificent barred spiral galaxy in the
southern constellation Hydra, is often called
the Southern Pinwheel Galaxy. It resembles the
Milky Way in overall form, but is considerably
smaller, spanning only 60,000 light-years.

Opposite **THE LOOSELY WOUND BARRED
SPIRAL NGC 925**
The unusual barred spiral galaxy NGC 925
lies in the constellation Triangulum, not far in
our sky from the Triangulum Galaxy, M33. This
highly inclined barred spiral features a disk
pocked with a few regions of star formation,
and lies some 30 million light-years distant. It
is a member of the NGC 1023 Galaxy Group,
a small association of at least five significant
members.

M83, A BEAUTIFUL FACE-ON BARRED SPIRAL
Often called out as a similar galaxy in structure
to the Milky Way, M83 is one of the most
beautiful galaxies in the southern sky, lying
in Hydra. As it lies only 15 million light-years
away, the detail in this galaxy is impressive,
with pinkish areas of hydrogen clouds and
clumps of star formation. M83 is about 55,000
light-years across, making it a roughly half-size
model of our galaxy. Its graceful form has led
to the sometimes-used nickname "Southern
Pinwheel Galaxy."

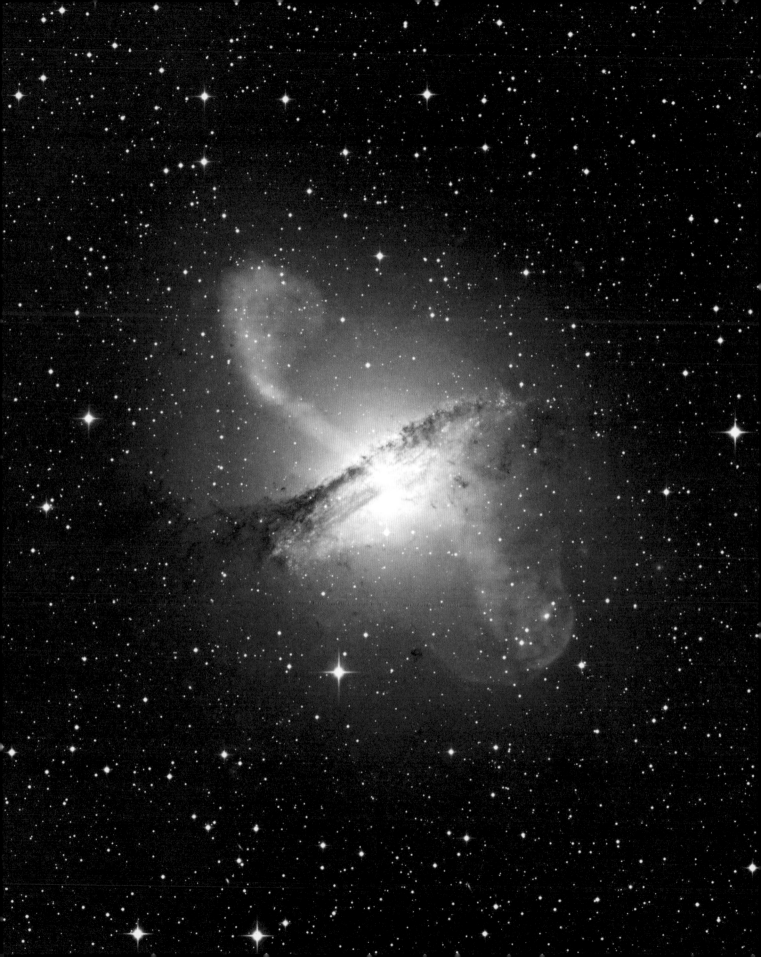

Opposite **THE COSMIC UPHEAVAL OF CENTAURUS A**
The weird galaxy Centaurus A (NGC 5128) in the southern sky is a classic example of a full-on galaxy merger. Two previous galaxies slammed together to form a highly energized, spherical mess that has unleashed a furious new round of star formation. This view from the Chandra X-ray Observatory provides a fresh look at the power of a galaxy's black hole. The X-ray jet at upper left, in orange, was depicted with the APEX Telescope in Chile, while the bluish X-ray data are from Chandra.

Right **NGC 4650A: A POLAR RING GALAXY**
The faint galaxy NGC 4650A In Centaurus is a strange creature—a so-called polar ring galaxy. This arrangement consists of a ring of material oriented around the galaxy's poles, which rotates around it. The main body of the galaxy is a lenticular-type object, meaning it's lens-shaped. The polar ring, oriented at right angles to the galaxy, is thought to have formed from an ancient collision. This weird object lies some 130 million light-years away.

Overleaf **NGC 2623: A GALACTIC TRAIN WRECK**
The highly distorted galaxy NGC 2623 in Cancer, lying some 250 million light-years away, previews what will happen to the Milky Way and Andromeda. This is a full-on merger of two galaxies that have lost their individual identities and now appear as one spherical mass with distorted, tidally disrupted tails. The sweeping curves of these tails stretch for more than 50,000 light-years and contain numerous clusters of hot, young stars.

193

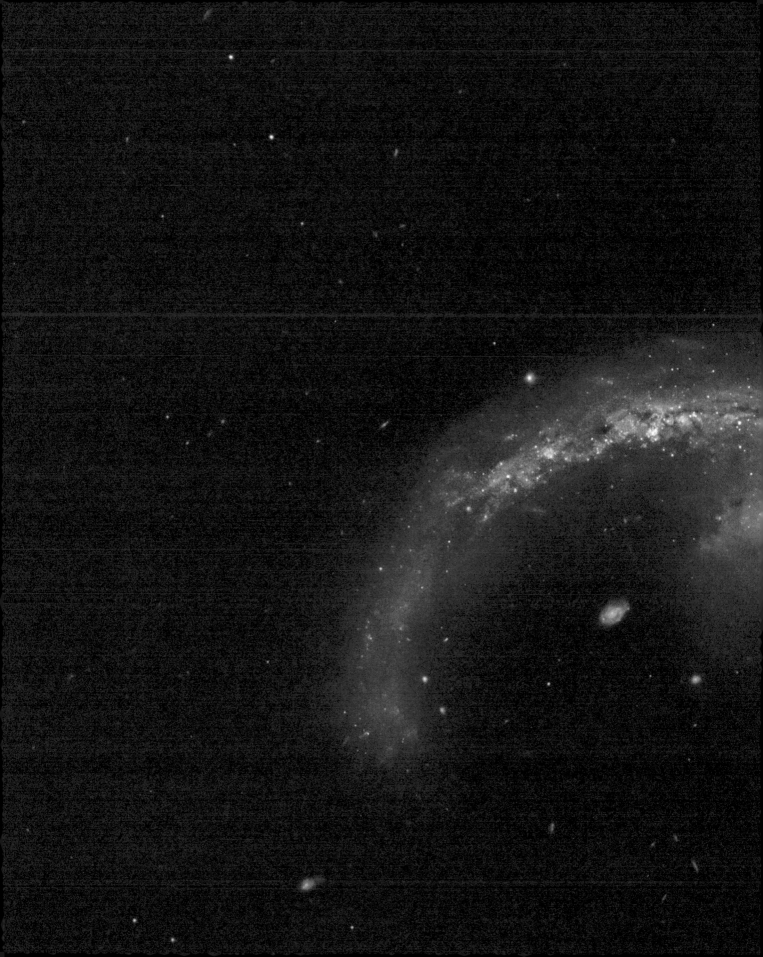

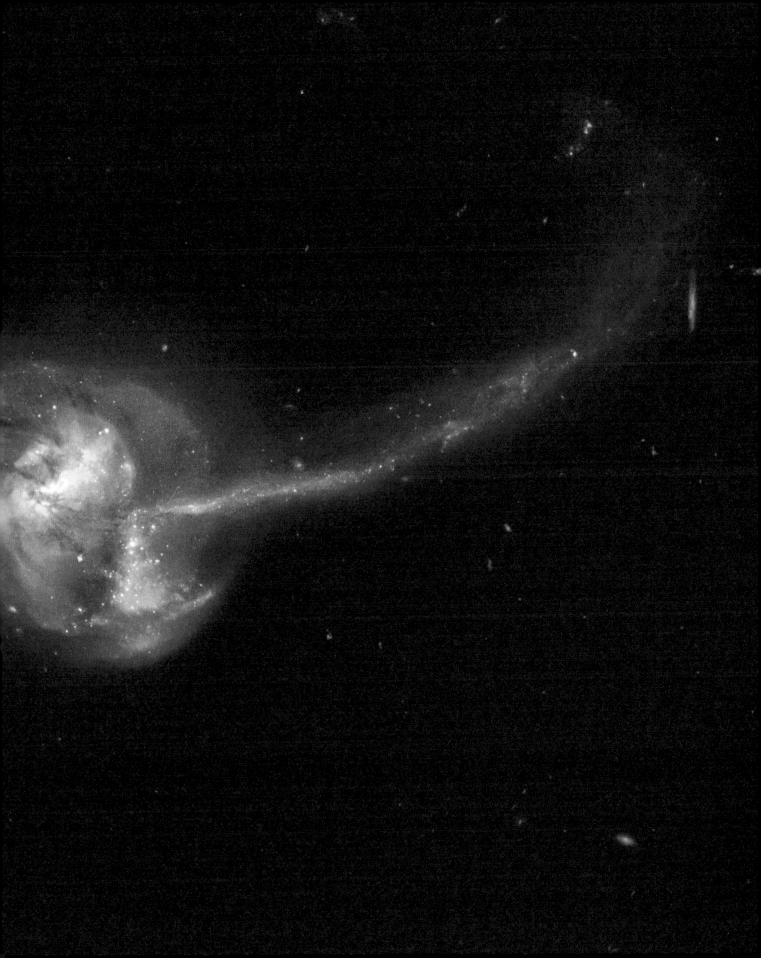

Opposite **NGC 5544 AND NGC 5545: GALAXIES "STAPLED TOGETHER"**

The interacting pair of galaxies NGC 5544 and NGC 5545 in Boötes appear as if they have simply been placed down on paper and stapled together. NGC 5544 is the bright, face-on barred spiral at right; its fainter companion juts straight into the side of the larger galaxy. The galaxies lie at a distance of 140 million light-years and give us a glimpse of what the early stages of the Milky Way–Andromeda merger might look like.

Above **THE ELEGANT INTERACTING GROUP STEPHAN'S QUINTET**

Located in Pegasus at a distance of more than 20 million light-years, Stephan's Quintet is one of the sky's premier groups of interacting galaxies. But it includes an impostor: the brightest galaxy, NGC 7320 (at right) lies only 39 million light-years away, and is projected onto the more distant galaxies. Other members are NGC 7319 (left of NGC 7320), NGC 7318A and NGC 7318B (above NGC 7320), NGC 7317 (above right), and NGC 7320C (lower left).

ONE OF THE SKY'S BEST: THE WHIRLPOOL GALAXY
The Whirlpool Galaxy in Canes Venatici, another
galaxy lying near the Big Dipper asterism in the
northern sky, is also known as M51 and is one of
the finest objects in the sky. An interacting pair of
galaxies, the Whirlpool, M51, is being passed by a
little interloper, NGC 5195, which is drawing material
off one of the larger galaxy's spiral arms. The
pair lie 23 million light-years away, and M51's disk
stretches across 60,000 light-years.

Opposite **THE ANTENNAE: A PICTURE OF THE FUTURE MILKY WAY?**

NGC 4038 and NGC 4039, collectively known as the Antennae galaxies, form an interacting pair of objects in the constellation Corvus, some 70 million light-years away. This kind of chaotic mish-mash, with the centers of two galaxies merging into one, offers a preview of what the future of the Milky Way might be, as the Andromeda Galaxy draws invariably closer. The collision between the Antennae began a shade less than a billion years ago.

Above **THE GRACEFUL EMBRACE OF THE ANTENNAE GALAXIES**

A photograph of the Antennae galaxies in Corvus, NGC 4038 and NGC 4039, reveals their dramatic appearance as viewed with large backyard telescopes. The shell-shaped bodies of the distorted galaxies are easy to spot: the graceful arcs of the faint tidal tails require photography to be seen.

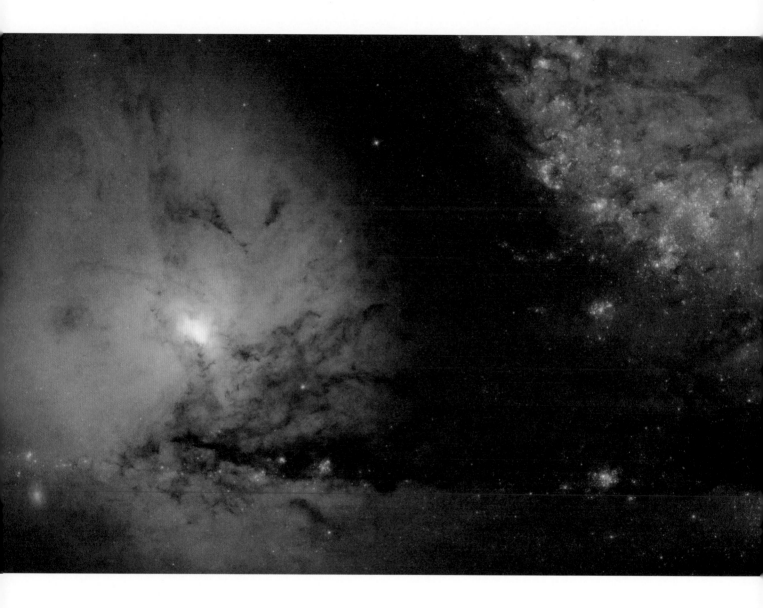

THE WHIRLPOOL'S INTERLOPER:
A CLOSE-UP ON NGC 5195

The famous Whirlpool Galaxy (M51) in Canes
Venatici is a favorite object for nearly all amateur
astronomers. It is an interacting system, and here
presented is a close-up of the little galaxy that
is passing it by, drawing material away from the
Whirlpool's disk (the edge is visible on the right-
hand side of the image). The system lies some
23 million light-years away, and the intricate dust
details and tight nucleus of this little galaxy span
20,000 light-years.

NGC 2207 AND IC 2163: A FULL GALACTIC EMBRACE

Two prominent galaxies in the constellation Canis Major reveal the early stages of a major, head-on collision. NGC 2207, on the left, is a barred spiral with a weak ring structure. IC 2163, a smaller barred spiral, will ultimately completely merge with the larger galaxy. These majestic objects lie some 80 million light-years away.

Above **THE PINWHEEL GALAXY SHINES IN ITS GLORY**

M101, also known as the Pinwheel Galaxy, is one of the brightest galaxies in Ursa Major, and lies near the Big Dipper asterism. This grand face-on spiral lies at a distance of 21 million light-years and is a monster spiral compared with the Milky Way. It stretches more than 170,000 light-years and contains a trillion stars.

Opposite **A DEEP SHOT REVEALS ANDROMEDA'S HALO STARS**

The Hubble Space Telescope produced this incredibly long exposure showing halo stars in the Andromeda Galaxy. Astronomers used the image to calculate that these stars formed only 6 to 8 billion years ago; they are younger than expected. The difference in ages in Andromeda's stars suggests a long history of mergers, in which younger stars mixed with older suns.

Opposite **M82: AN EXPLODING "CIGAR GALAXY" THAT DELIGHTS OBSERVERS**
The edge-on galaxy M82 is a marvel of the northern sky: Lying next to M81, also a bright galaxy, it is easily visible in backyard telescopes in the constellation Ursa Major. M82 lies some 12 million light-years away, and is a so-called starburst galaxy, undergoing a huge episode of star formation, and spewing material outward from its violently engaged nucleus. Gravitational forces from the nearby galaxy M81 are causing the starburst event. The galaxy's core holds a supermassive black hole.

Above **NGC 3370: A GRAND DESIGN SPIRAL GALAXY**
This stately spiral in the constellation Leo, which lies some 100 million light-years distant, has features that resemble our galaxy—a so-called grand design, showing prominent and well-defined spiral arms. This Hubble Space Telescope image has enabled astronomers to study individual stars within NGC 3370, leading to accurate distance measurements and an understanding of how this galaxy works.

Overleaf **NGC 4945: A STATELY SOUTHERN EDGE-ON SPIRAL**
Seen at a nearly edge-on angle to our line of sight, spiral galaxy NGC 4945 in the southern constellation Centaurus is about the same size as the Milky Way. The galaxy's peculiar center is that of a Seyfert galaxy, with a high-energy nucleus, and harbors an active black hole that is spewing out material at high speeds.

Chapter Five

GALAXIES TO THE EDGE OF THE UNIVERSE

✳ ✳ ✳

Now we're getting to know just how vast the cosmos really is. Consider our spaceship of the imagination. Traveling as fast as a photon, at the speed of light, it took us 100,000 years to cross our galaxy's disk and 10 million years to go from one side of our Local Group to the other. And to travel to the core of the Virgo Cluster? More than 50 million years. Now consider how long it takes to get from galaxy cluster to galaxy cluster, or to the most distant galaxies we can see. Now we're talking about billions of years of travel time even at the fastest speed in the universe. With a simple backyard telescope, you can see galaxies and quasars that are several billion light-years away. Some of those photons striking your eye left their home galaxy before the Sun and Earth existed.

FAST FACT
The present-day universe contains some 100 billion galaxies. Early in its history, it contained probably more than 1 trillion, but many smaller galaxies have since merged.

As we move farther out into the universe, we see many unusual galaxies and groups of galaxies along the way that demonstrate the many types.

To grasp the size of the universe at very large scales, we need to build on stepping-stones. Most of the stars and galaxies we see in the sky are so distant that we don't see them move from night to night—they're so far away that in effect we see the universe as one frame of a giant cosmic movie. So for a moment let's consider things that for humans may be very far away but from a cosmic perspective are really close.

Imagine the distance between Earth and the Sun, which astronomers call one astronomical unit, being 1 centimeter. On that scale, you can sketch out the solar system on several sheets of paper taped together, end to end. Make Mars 1.5 centimeters from the Sun and put Jupiter at 5 centimeters, Saturn at 9.5 centimeters, Uranus at 19 centimeters, and Neptune at 30 centimeters. Pluto and its fellow icy bodies lie some 40 centimeters out from the Sun, and many small asteroids are scattered farther away. But on this scale, the Oort Cloud, the physical outer edge of the solar system, which contains 2 trillion comets, lies 1,000 meters away—more than the length of 10 American football fields. And remember that humans, on this scale, have physically traveled as far away as the Moon—a tiny fraction of that first centimeter.

Even our own solar system is dauntingly large when considered as a scale model. Now consider that on this scale the Milky Way Galaxy would be unthinkably larger. You could stack more than 63,000 astronomical units side by side in one light-year, and the bright disk of the Milky Way is 100,000 light-years across. Those dreams of rocketing across the galaxy to visit other civilizations? Forget them. Leave cosmic travel to the movies.

Based on calculating the expansion of the universe backwards in time, and also understanding the nature of cosmology, astronomers now estimate that the universe overall is at least 93 billion light-years across. That may not seem right, given that the speed of light is fixed as the fastest speed anything can move and that the universe is 13.8 billion years old. But again, remember that space itself expands over time. There's another catch as well. The 93 billion light-years that astronomers estimate as the size of the cosmos refers only to the visible universe—the part we can see.

FAST FACT

The earliest known galaxy thus far observed in the universe (and the most distant) is GN-z11, a blob of faint light that represents a protogalaxy some 13.4 billion years old. The object formed just 400 million years after the Big Bang, around the earliest time that galaxies could have come together.

COSMIC INFLATION
AND THE MULTIVERSE

In the 1980s, two cosmologists independently pushed forward an idea about the early universe. The American Alan Guth and the Russian-American Andrei Linde proposed concepts that came to be known as inflation theory. In short, if the very young universe hyperinflated almost instantly within the first second after the Big Bang—from the size of a pea to the size of a softball—then some aspects of what astronomers observe in the later universe can be explained pretty conveniently. Thus, most cosmologists now have solid confidence in inflation theory, and if it continues to hold up, then one of its implications is that the visible universe we see may not be the entire universe. In fact, as counterintuitive as it sounds, the universe could even be infinite. And other universes, the so-called multiverse, might exist beyond our own.

HOW BIG IS THE UNIVERSE?

To answer the question of how big the universe is, let's work with what we know—the visible universe. Of its diameter of 93 billion light-years, we have thus far explored our own Local Group of galaxies, the nearby clusters and groups of galaxies out to several tens of millions of light-years away, and the Virgo Cluster and Supercluster, which consists of about 100 galaxy groups and clusters and has a diameter of about 110 million light-years. So looking out all that way to consider some of the most distant galaxies that backyard observers armed with small telescopes can observe, we are still just scratching the surface. The diameter of the Virgo Supercluster represents about one-one-thousandth of the diameter of the entire visible universe. Some of the notable nearby clusters include the Hercules Galaxy Cluster, the Pegasus Galaxy Cluster, and the Corona Borealis Galaxy Cluster.

So what in the world exists out beyond the Virgo Supercluster? The bulk of the universe far beyond most of what we can easily see from Earth must be loaded with all kinds of interesting and amazing things. Don't you think so? If you do, you are exactly right.

THE LARGE-SCALE UNIVERSE

The last generation has been studded with discoveries that clarify the murky picture of what the universe is like on very large scales. Progress really began in earnest in the 1970s, when astronomers pushed large sky survey projects forward. Structures even larger than superclusters began to appear, and astronomers dubbed them sheets, walls, and filaments—huge areas of galaxy superclusters winding their way around the universe, separated by enormous voids or chasms. Cosmologists like to imagine the very large-scale structure of the cosmos as a giant foam, with bubbles representing threads of galaxies and the insides of bubbles the voids separating them. You could also imagine a complex three-dimensional spiderweb, with the woven web representing filaments of galaxies and the intervening cavities the voids separating them.

Beginning in the 1980s, continued surveying of galaxies in great numbers added enormously to our picture of the large-scale universe. Brent Tully and his collaborators mapped out the Pisces-Cetus Supercluster Complex, the huge wall of galaxies that contains the Virgo Supercluster (including the Local Group and the Milky Way). About the same time, astronomers identified a giant cavity in the local universe called the Giant Void.

DISTANT GALAXY CLUSTERS

Beyond the Virgo Supercluster lie numerous clusters and superclusters of galaxies, perhaps some 10 million superclusters altogether—and remember, we are still only in the observable universe. Collectively, astronomers refer to this universe on a macro scale as the large-scale structure of the cosmos. Telescopes are time machines; as we look at more distant objects, we're seeing them as they were long ago. Since 2012, astronomers have studied the so-called Hubble

FAST FACT
The most active galaxy known is EQ J1000054+023435, sometimes called the Baby Boom Galaxy, which lies at a distance of 12.2 billion light-years. It produces 4,000 new stars each year, as compared with the Milky Way's output of about 10 stars per year.

THE BIG BANG

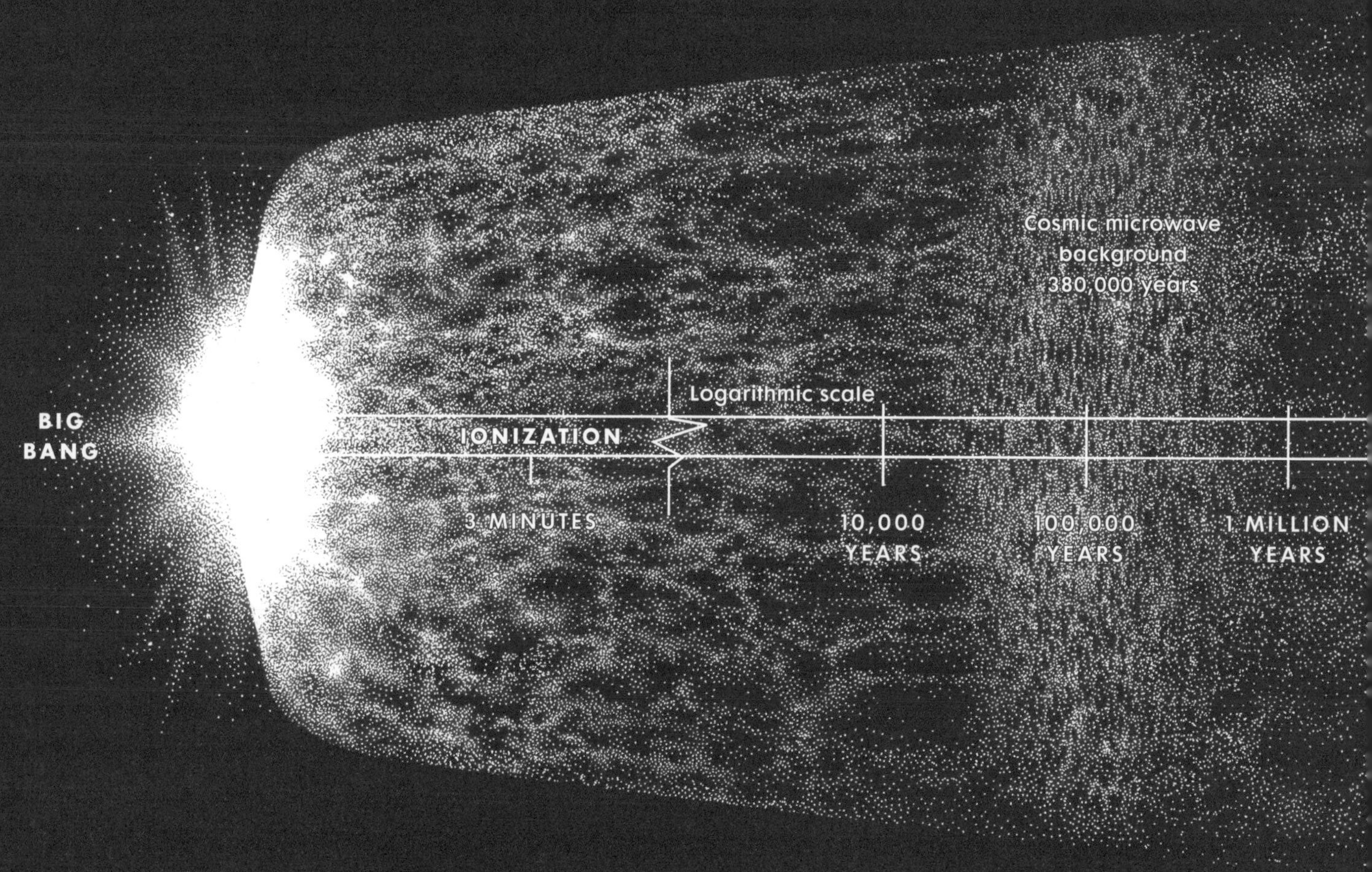

BIG BANG

IONIZATION

Logarithmic scale

3 MINUTES

10,000 YEARS

100,000 YEARS

1 MILLION YEARS

Cosmic microwave background 380,000 years

Astronomers think a span of time that lasted hundreds of millions of years called the reionization era began when the first stars and galaxies emitted radiation that turned hydrogen atoms (one proton, one electron) into hydrogen ions (a proton without an electron). Thus began the era of galaxies, and moving this diagram to the right brings us to the present day.

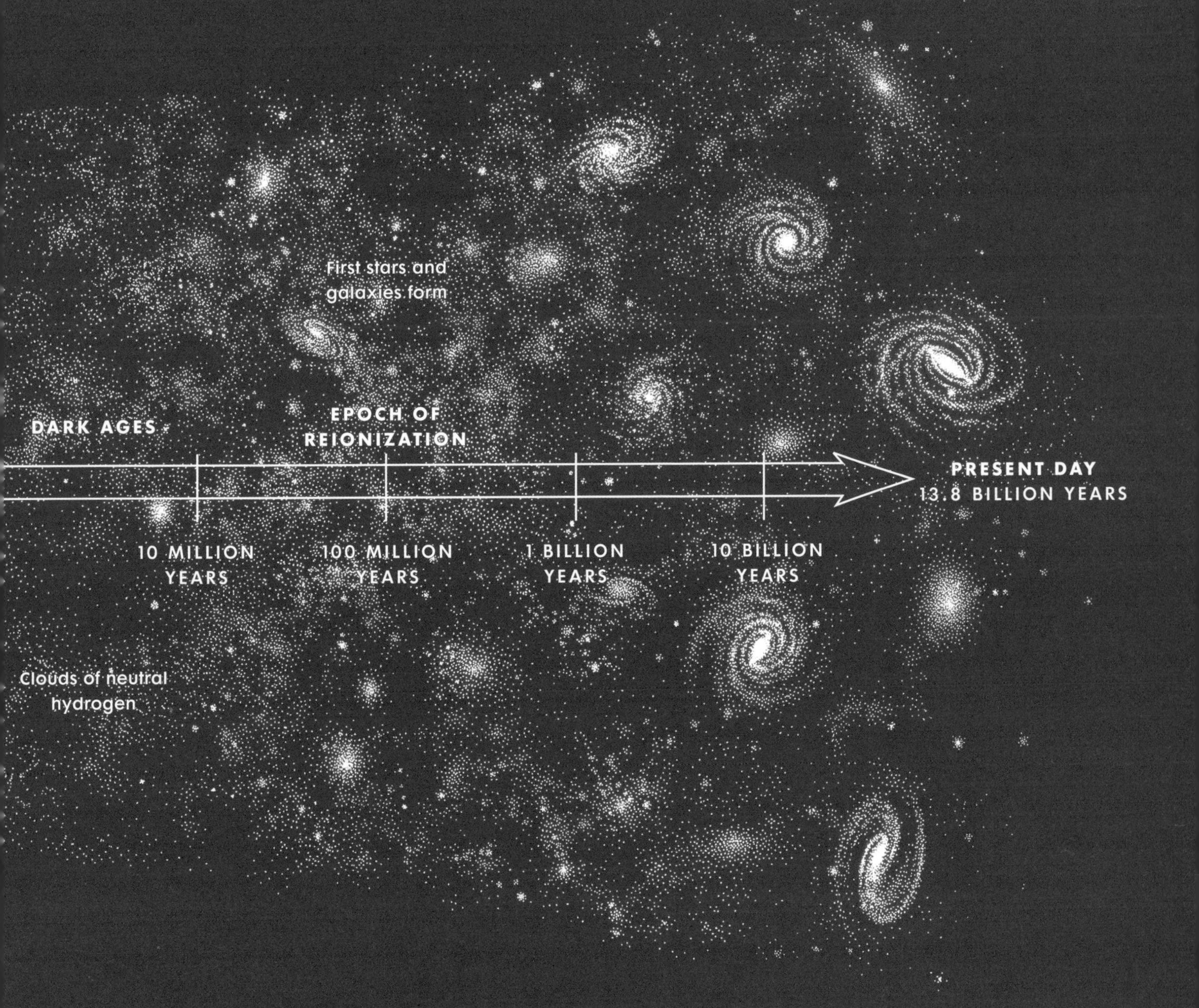

First stars and
galaxies form

DARK AGES

EPOCH OF
REIONIZATION

PRESENT DAY
13.8 BILLION YEARS

10 MILLION
YEARS

100 MILLION
YEARS

1 BILLION
YEARS

10 BILLION
YEARS

Clouds of neutral
hydrogen

eXtreme Deep Field, the field of incredibly distant galaxies imaged for a long period, in the direction of the constellation Fornax. Here they see primitive, bluish protogalaxies that formed some 13.2 billion years ago, only 600 million years after the Big Bang. These primitive galaxies are the seedlings that came together by gravity to make normal galaxies as time rolled on. By contrast, the most remote mature cluster of galaxies known, CL J1449+0856, lies more than a third of a universe away from us on that scale of the universe's diameter of 93 billion light-years.

Some galaxy clusters and superclusters are very important laboratories that help astronomers understand the nature of dark matter, the unseen material that makes up a significant fraction of the universe. These include Abell 520, the Bullet Cluster, Dragonfly 44, El Gordo, MACS J1206.2-0847, and Pandora's Cluster.

ALONG COMES THE GREAT WALL

In the last generation, astronomers explored a variety of relatively close galaxy superclusters. These included the Coma Supercluster (20 million light-years away), the Perseus-Pisces Supercluster (100 million light-years), the Hercules Supercluster (330 million light-years), and the Shapley Supercluster (400 million light-years). Lying in the direction of the constellation Centaurus in our sky, the Shapley Supercluster contains the largest concentration of galaxies in our local region of the universe. The American astronomer Harlow Shapley first reported an unusual number of distant galaxies in this region in 1930; decades later, astronomers recognized the area as hosting a supercluster and named it for the great astronomer.

In this last generation, a picture of the local universe started to come together. By the end of the 1980s, a team of astronomers at the Harvard-Smithsonian Center for Astrophysics, including Margaret Geller and John Huchra, had identified the so-called Great Wall, an enormous sheet of galaxies stretching 500 million by

200 million by 15 million light-years across. An important survey project called the Sloan Digital Sky Survey, which commenced in 2000, three years later identified another mammoth structure, the Sloan Great Wall. Announced by the American astronomer J. Richard Gott and his collaborators—and again the discovery involved Geller and Huchra—the Sloan Great Wall is at least twice the size of the Great Wall first announced by Geller and Huchra, stretching some 1.4 billion light-years across.

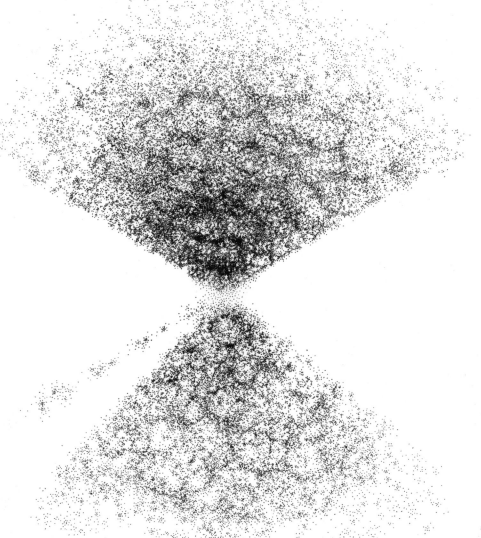

LANIAKEA

ENTER THE GREAT ATTRACTOR

One curious finding from the surveys beginning in the 1970s was an anomaly in the expansion of the universe. Through multiple observations, astronomers noticed that some large mass seemed to be tugging on the local universe, pulling us and the galaxies near us toward a point in the direction of the Southern Hemisphere constellations Triangulum Australe and Norma. This puzzled astronomers for a long time, even as they dubbed the gravitational oddity the "Great Attractor." This mass of galaxies that is pulling us toward it lies some 200 million light-years away.

In the last twenty years, the story of the Great Attractor has become even more complex. Using sophisticated X-ray observations, astronomers refined their understanding of the effect of the Great Attractor on us, demoting it in importance. A mass of galaxies is indeed exerting a gravitational force on us in this direction, but its effect is not as significant as astronomers first believed. Now astronomers believe that the galaxies in our neighborhood are being pulled toward a larger, more massive structure lying farther away than the Great Attractor—the Shapley Supercluster. And astronomers have also found that the local universe belongs to a more newly discovered supercluster, only detected in the last few years.

FAST FACT

The galaxy ESO 137-001, lying some 220 million light-years away in the constellation Triangulum Australe, has a long tidal tail that features the highest rate of star formation known outside a galaxy's main body.

A SURPRISING DISCOVERY: LANIAKEA

Once again, Brent Tully and his team reset knowledge of the local universe, this time in 2014, with his identification of a new supercluster based on understanding the relative motions of galaxies in a more sophisticated way than ever before. Mapping the local structures out more thoroughly than ever before, Tully and his colleagues identified what they called the Laniakea Supercluster—named for the Hawaiian word meaning "immense heaven."

The Laniakea Supercluster, also sometimes now called the Local Supercluster, contains about 100,000 of the galaxies closest to us in the universe, including the Local Group and the Milky Way. Although this massive cluster is now traveling together through space, not all of the galaxies within it are gravitationally bound. Eventually, at least part of Laniakea will splinter off.

Astronomers believe that the overall diameter of Laniakea is about 520 million light-years. It contains the mass of 100,000 Milky Way galaxies and has four major components. They are the Virgo Supercluster, which contains almost all of the bright galaxies in our sky, including the Local Group and the Milky Way; the Hydra-Centaurus Supercluster, which includes the Great Attractor, the Hydra Supercluster or Antlia Wall, and the Centaurus Supercluster; the Pavo-Indus Supercluster; and the Southern Supercluster.

DISTANT GALAXY
SUPERCLUSTERS

We have visited the Virgo Supercluster in some detail. But what about the others?

Surrounding Laniakea in the local universe are several other superclusters—the Shapley Supercluster, Hercules Supercluster, Coma Supercluster, and Perseus-Pisces Supercluster. Each of these structures contains hundreds of galaxy clusters and groups and is linked in the structural web of the cosmos, separated by vast voids. Galaxies exist in groups, chains, and filaments, and where they don't exist, there is only vast empty space, a staggeringly wide chasm of darkness.

Galaxy clusters also provide a way to see objects far behind them. Their gravity can act as a lens, focusing the light from really distant galaxies and quasars into images or arcs that can be studied. There are many great examples, including Abell 1689, the Cheshire Cat, SDSS J1531+3414, and MACS J1149.6+2223.

FAST FACT
In addition to GN-z11, other galaxies that are among the most distant are MACS1149-JD (13.3 billion light-years away), EGSY8p7 (13.2 billion light-years), and EGS-zs8-1 (13.1 billion light-years).

THE LARGEST STRUCTURES
IN THE UNIVERSE

All of this thinking about the universe at really large scales pushes astronomers back into thinking about how matter is organized in the cosmos. On a relatively small scale, matter is organized into stars, and stars into galaxies. We have seen that galaxies are organized into groups, clusters, superclusters, walls, sheets, and filaments, on increasingly larger scales. Throughout the initial major galaxy surveys, astronomers believed that superclusters were the largest structures that exist. By the early 1980s, however, they began to find evidence for even larger structures. Objects called large quasar groups (LQG) baffled astronomers at first. In 1982, the Scottish astronomer Adrian Webster found what came to be called the Webster Large Quasar Group, a collection of five quasars stretching over 330 million light-years. Recall that quasars are the highly energetic centers of young galaxies, driven to enormous bouts of activity by central supermassive black holes.

Nearly two dozen LQGs are now known, and they are thought to be some of the largest structures in the cosmos. The so-called Huge LQG, discovered in 2013, contains seventy-three quasars spread over a diameter of 4 billion light-years.

The Clowes-Campusano Large Quasar Group contains thirty-four quasars throughout a structure that measures some 2 billion light-years across. Discovered by the English astronomer Roger Clowes and the Chilean astronomer Luis Campusano in 1991, this massive object lies some 9.5 billion light-years away toward the constellation Leo, and not terribly far from the so-called Huge LQG. It's possible that the two structures are related.

Another large quasar group, U1.11, is larger still. Located in our sky toward the constellations Leo and Virgo, this odd group contains thirty-eight quasars in a region spanning 2.2 billion light-years. The fact that young, highly energetic galaxies are emitting copious amounts of radiation as quasars, concentrated in a region such as this, suggests that large quasar groups like U1.11 are signaling the formation of a galaxy filament.

Objects called Large Quasar Groups baffled astronomers at first.

The previously mentioned Huge LQG offers astronomers not only insights but a bit of a battleground as well. With some seventy-three quasars within a structure with a diameter of some 4 billion light-years, Huge LQG seems to be a massive formation within the large-scale cosmos. Clowes reported the finding in 2013, and the structure was studied by numerous other astronomers in the years following. Strangely, Clowes reported that the Huge LQG appears to violate the cosmological principle—that is, the idea that at very large scales, the universe is homogeneous, or relatively smooth and uniform. The lumpiness in this structure challenges this notion, although astronomers have debated the definition and whether it really causes a problem. Some astronomers have also disputed the existence of this structure, but Clowes and other researchers have offered further evidence for its existence.

THE HERCULES-CORONA BOREALIS GREAT WALL

In late 2013, a team of American and Hungarian astronomers identified the Hercules-Corona Borealis Great Wall, sometimes called the Great GRB Wall, a massive structure found by identifying gamma-ray bursts (GRBs) in this particular region of sky. Gamma-ray bursts are extraordinarily energetic events that have been observed in distant galaxies and may result from the explosion of a supernova or a hypernova, ending the life of a massive star. The rapidly rotating, dying star may form a neutron star, quark star, or black hole, and flash an incredible burst of energy as it does so. Astronomers have observed far more of these bursts in a particular region of sky than is statistically predicted, suggesting the existence of this rich structure of galaxies that stretches as much as 10 billion light-years across.

If the Hercules-Corona Borealis Great Wall exists, it is the largest known structure in the universe. Other large structures offer unusual insights into the cosmos as a whole. Brent Tully's 1987 discovery of the Pisces-Cetus Supercluster Complex helped to sort out the large structure around our own Virgo Supercluster, which it contains. Pisces-Cetus is a large galaxy filament that stretches about 1 billion light-years long and 150 million light-years wide and contains some sixty galaxy clusters. This complex contains five major pieces: the Pisces-Cetus Supercluster, the Perseus-Pegasus Chain, the Pegasus-Pisces Chain, the Sculptor Region (including the Sculptor Supercluster and the Hercules Supercluster), and the Laniakea Supercluster, which contains the Virgo Super-cluster (and us!) and the Hydra-Centaurus Supercluster.

Many of these large structures contain strange kinds of galaxies that astronomers are just beginning to understand. These include objects like gamma-ray bursts, double quasars, distorted tails, galaxies that eject stars, and extremely faint galaxies.

> If the Hercules-Corona Borealis Great Wall exists, it is the largest known structure in the universe.

The universe is truly so big that it is difficult to comprehend. We can think about the scale of our solar system quite clearly: we can measure out the distances between the Sun and the planets on a simple sheet of paper, and we can even clearly envision the physical edge of the solar system at the enormous distance to the Oort Cloud. By association, it's easy to imagine the size of our own Milky Way Galaxy.

But the scale of the cosmos that carries us out to the galaxies surrounding ours—to the Virgo Cluster, to our own supercluster of galaxies, and beyond—staggers the human mind. On the one hand, the enormity of the universe makes us feel small and insignificant, living briefly on Earth in an unbelievably small corner of the cosmos. And yet the fact that we are made of the stuff of the universe and are sentient—that we can think and look out to the stars and reflect on the meaning of it all—gives us an amazing power that is also almost beyond belief.

BLACK HOLES:
UBIQUITOUS IN THE COSMOS

So what exactly do black holes look like anyway? Black holes come in many sizes, at least presumably. Stellar black holes like Cygnus X-1 have more than about five times the mass of the Sun. And yet they are small, physically only about 20 kilometers across, and of course they are black. They are so small that within the distance

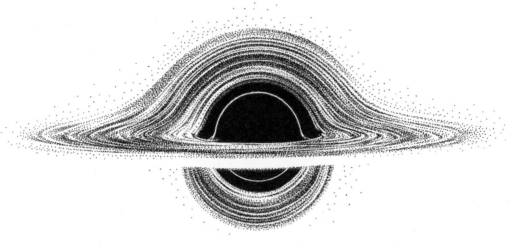

This simulation gives a realistic depiction of a black hole accretion disk, including the light-bending effects of relativity.

FAST FACT
The galaxy with the most massive black hole known is TON 618, a quasar 3.2 billion light-years away in Canes Venatici. Its central black hole "weighs" some 66 billion solar masses.

of the nearest star to the Sun, 4 light-years, we would not be able to detect a stellar black hole unless it were interacting with a companion star, which would show distortion effects. Astronomers know of about two dozen stellar black holes in the Milky Way, and certainly numerous others exist.

Stellar black holes are classified into several types. All stellar black holes pull material into their dense centers and lock it away from the cosmos. Some rotate, however, and others don't. A Schwarzschild black hole is static, with no spin or magnetic field. This class was named for the German physicist and astronomer Karl Schwarzschild, a friend of Einstein's, who studied general relativity and tragically died young just after his service in World War I. A Kerr black hole, by contrast, rotates, possessing both spin and magnetic field. This class was named for the New Zealand mathematician Roy Kerr, another student of general relativity. A black hole with no spin but with a magnetic field is called a Reissner-Nordström black hole, after the German and Finnish physicists who studied them.

There should also be black holes of much smaller mass than stellar black holes, like mini or micro black holes. To turn Earth into a black hole, for example, you would need to squeeze it down to the size of a grape. And so-called intermediate-mass black holes, which have masses between those of stellar black holes and supermassive black holes, must also exist. These range from 100 to 1 million times the mass of the Sun. It's the supermassive black holes that dominate the centers of galaxies.

Supermassive black holes exist within the range of hundreds of thousands of solar masses to billions of solar masses. And yet they are physically only about as big as our solar system. They also have a spin and a magnetic field. An important 2013 study by the astronomers John Kormendy and Luis C. Ho lists eighty-five galaxies that show evidence of central supermassive black holes, based on dynamical modeling of the motions of objects around their centers. Using these data, they found a correlation between the mass of a central black hole and the host galaxy's central bulge, the brightest area surrounding the nucleus. They suggest that a galaxy's central black hole and its central bulge grow together as they evolve, the bulge receiving some of the material that slingshots around the black hole instead of falling in.

Because supermassive black holes are physically much larger than stellar black holes, the time scales of their activity are much longer too. The cosmic snapshot we see of them in the universe right now has varied over time. Many supermassive black holes go through phases of intense activity when matter happens to fall into them and they become briefly active; then they hibernate for long periods. They can be seen in action only when caught at the right time. We can see this in many galaxies, such as M74, M82, NGC 1032, and NGC 6240.

The parts of a black hole illuminate just how these strange creatures work. The center of the black hole is marked by the "singularity," where matter is infinitely dense and space-time is infinitely curved, or warped. The physics we're used to break down in the singularity. Anything that falls into the singularity would be crushed and added to the mass of the black hole. Before that point, there is an "event horizon"—the boundary in space-time beyond which light and matter falling into the black hole cannot escape.

FAST FACT
Of all the galaxies that backyard astronomers can observe, the one with the most massive black hole is NGC 4889 in the Coma Cluster of galaxies. Its supermassive black hole weighs in at about 20 billion solar masses.

A rotating black hole also has an ergosphere, a region just outside the black hole in which so-called frame dragging occurs and where space-time is dragged faster than the speed of light. If you approached a rotating black hole, you wouldn't see this, but the effect twists space-time around the black hole like cake batter flying around a beater in a bowl. The word "ergosphere" is derived from the Greek *ergon*, for work, because it is possible to extract energy and mass from this region. Surrounding the black hole and the ergosphere is a photon sphere, a region in which photons travel around the black hole in unstable circular orbits instead of the straight paths on which they normally travel. This creates a "shadow" outlining the shape of the black hole.

Further out, an accretion disk forms around the black hole. This disk consists of matter in an omnidirectional halo that spirals around and gradually into the black hole. Jets of matter can also fling away from the black hole or blast away from its poles at near light speed.

One of the great questions of science fiction is, of course, what would happen to you if you fell into a black hole? The answer depends on the kind of black hole we're talking about. Smaller black holes are actually more lethal. If you fell feet-first into the event horizon of a black hole with a mass ten times that of the Sun, you would be stretched vertically, squashed sideways, and pulled into a string of particles— "spaghettified." It would do you absolutely no good at all. End of story.

However, if you fell into a supermassive black hole at the center of a galaxy, the situation would be different—hypothetically, of course. The gravitational forces of a million-solar-mass black hole are very different, and they would allow you to reach the event horizon safely. In fact, you would probably never know that you had passed through the event horizon. Companions watching you from the outside would not notice you crossing it either. They would see you slow down and "hover" outside it, becoming dimmer and redder to their vision until you disappeared. But then you would be gone from your companions forever, and ultimately crushed by the singularity.

> What would happen to you if you fell into a black hole? The answer depends on the kind of black hole.

FAST FACT
In 2019 astronomers imaged the emission around the black hole in M87, a black hole that weighs in at 6 billion times the mass of the Sun.

THE CONCEPT
OF TIME TRAVEL

The romance of traveling through a black hole accelerated from a subject of academic discussions into a sci-fi staple after the American theoretical physicist John Archibald Wheeler coined the term "wormhole." Postulating that some black holes might offer temporary "tunnels" allowing travel shortcuts from one time and place in the universe to another, he instigated a cottage industry of fun.

But Stephen Hawking reminded us that in a practical sense, time travel may not work. "Whenever one tries to make a time machine," writes Hawking, "and no matter what kind of device one uses in one's attempt (a wormhole, a spinning cylinder, a 'cosmic string,' or whatever), just before one's device becomes a time machine, a beam of vacuum fluctuations will circulate through the device and destroy it."

MARKARIAN 231

Quasar with a double black hole

The intensity of light from quasars is remarkable: they
are emitting tremendous amounts of radiation cast
outward from their black hole accretion disks. The
closest quasar to Earth is Markarian 231, a galaxy
580 million light-years away in Ursa Major. This artwork
depicts the double black hole that lies at the heart of
Markarian 231, which creates a doughnut-shaped disk.

Today the project of detecting the black hole engines that drive galaxies is in a whole new phase. In the words of Kip Thorne, colliding black holes are "the most luminous objects in the universe—but no light!" What he means is that when black holes merge, when they are drawn together by immense gravitational pulls and clash and collide in galactic battle, they should emit enormous numbers of gravitational waves—ripples in the curvature of space-time that would propagate across the cosmos and be detectable with special instruments.

In 1992, Thorne and his associates founded the Laser Interferometer Gravitational-Wave Observatory (LIGO), a complex physics experiment with facilities in Hanford, Washington, and Livingston, Louisiana. Designed to detect gravitational waves, this experiment had an historic success in early 2016, when LIGO scientists announced the first verified gravitational wave detection from the collision of two black holes. This was a huge deal and another verification among many that Einstein's theories of relativity and predictions of black holes were right.

The actual detection took place in September 2015, when a gravitational wave produced by the merger of two distant black holes swept past Earth. The black holes had approximate masses of thirty-six and twenty-nine times that of the Sun and were located some 1.3 billion light-years away. The "chirp signal" that made up the detection lasted for 0.2 second. LIGO has since made more detections, and others will undoubtedly follow.

The evolution in our thinking about black holes has been a long time coming. When I joined the staff of *Astronomy* magazine in 1982, black holes were largely a rumor. Long postulated and believed to exist since the early 1970s, they didn't receive ironclad credence until 1990. By the end of that decade, not only had astronomers begun to appreciate how widespread black holes are as the driving "central engines" of most galaxies, but they had also begun to understand that galaxies were far more violent and energetic in the early days of the cosmos. They were starting to see patterns emerge of galaxy behavior over billions of years of cosmic time. Galaxies, like people, they found, evolve, changing their character in important and dramatic ways as the years roll on.

Opposite **ARP 248: THE DISTANT GLOW OF WILD'S TRIPLET**
This group of three faint galaxies in the constellation Virgo lies at a distance of 225 million light-years and offers a still frame of a prolonged cosmic dance. The tidal tails from the brightest galaxy form an apparent bridge to the edge of the other two galaxies' disks.

Above **ESO 243–49 AND ITS HYPERLUMINOUS X-RAY SOURCE**
The strange edge-on galaxy ESO 243–49 lies at a distance of 290 million light-years in the constellation Phoenix. Within this galaxy, astronomers have discovered what they term a hyperluminous X-ray source, HLX-1 (circled), which they believe is a strong candidate for an unusual type of black hole, an intermediate-mass black hole. The intensity of the X-rays and spectrum of the object indicate a young, hot cluster of blue stars some 250 light-years across surrounding this black hole. The mass of the black hole is believed to be 100 to 100,000 solar masses.

Top MONSTER GALAXY POWERED BY BLACK HOLE

A giant elliptical galaxy designated A2261–BCG lies at the heart of Abell 2261, a galaxy cluster in Hercules some 3 billion light-years away. It has the largest core region of a galaxy yet observed, stretching some 10,000 light-years across, and is powered by a supermassive black hole pair that may have stirred up star formation in the galaxy's central region.

Bottom AM 0644–741: A GALACTIC TRAIN WRECK WITH RINGS

This very strange galaxy, designated AM 0644–741, lies at a distance of 300 million light-years away in the southern constellation Volans. It is a prime example of a polar ring galaxy, in which a high-speed intruder (not visible in the image) punched head-on through the galaxy in a cosmic train wreck. The collision caused the glowing ring of bluish stars and gas surrounding the inner, yellowish core. The ring stretches 130,000 light-years across, larger than the diameter of the Milky Way.

Previous **INFANT STARS LITTER THE SMALL MAGELLANIC CLOUD**

Embryonic stars still in formation appear wrapped in a bluish coating of nebulous gas in this Hubble Space Telescope image of a portion of the Small Magellanic Cloud. The nebula, designated NGC 346, is the brightest in the satellite galaxy, and consists of gravitationally collapsing gas clouds.

Opposite **FORNAX A: A COSMIC DUST BUNNY**

The strange distorted galaxy NGC 1316, known as Fornax A because it is a radio source, consists of a highly chaotic barred spiral galaxy that looks more like an elliptical. The galaxy appears to have been built up by repeated mergers over distant cosmic time, some 3 billion years ago. The strong radio emission comes from a powerful central black hole. The dusty lanes circling this galaxy are witness to the disruptive forces within it. Fornax A lies 62 million light-years away.

Top **THE BIRTH OF A TINY GALAXY**

The incredible power of the Hubble Space Telescope comes through via this image of POX 186, a tiny dwarf galaxy in Virgo, lying at a distance of 69 million light-years. This unusual object is a blue compact dwarf galaxy, an unusual class that features hot blue stars and tiny diameters. POX 186 spans a mere 900 light-years, less than 1 percent the size of the Milky Way. The fact that this galaxy is so young, is just now forming, suggests that some late-blooming, small galaxies may be the last to form in a succession of the universe's timeline.

Bottom **A SPIRAL GALAXY EMITS A POWERFUL JET**

Nearly all galaxies with incredibly powerful central black holes, spewing jets angrily out from their centers, are huge elliptical galaxies. Astronomers using the Hubble Space Telescope imaged a spiral galaxy, 0313–192, emitting such a powerful jet. This galaxy, one billion light-years away in Eridanus, gives astronomers a new type of object to study.

Opposite **A POCKET OF STAR FORMATION IN THE LARGE MAGELLANIC CLOUD**
The region of swirling gas and dust designated LH 95 lies within the Large Magellanic Cloud, the Milky Way satellite galaxy some 163,000 light-years away. A shroud of bluish haze covers massive, newborn suns along with low-mass infant stars.

Above **NGC 4214: A DWARF GALAXY ABLAZE WITH STARS AND GAS**
The dwarf irregular galaxy NGC 4214 in Canes Venatici lies a mere 10 million light-years away, making it close enough to record substantial details in. A larger, brighter version of the Small Magellanic Cloud, the galaxy contains a laboratory of bright pinkish gas clouds, dazzling blue star clusters, and other evidence of new rounds of star formation.

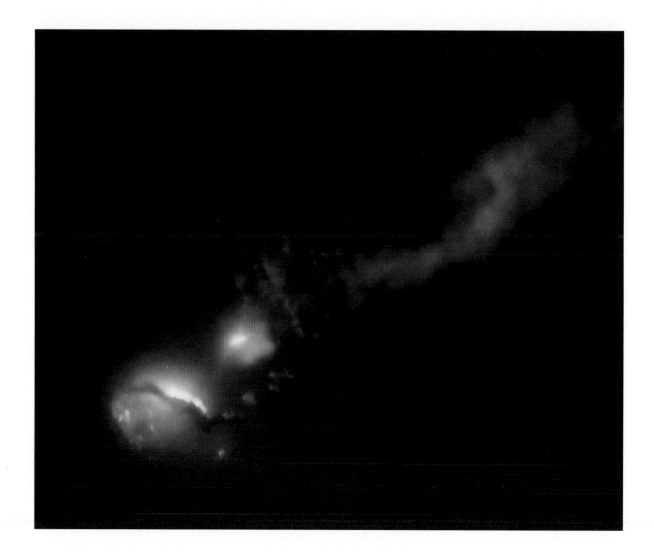

Above **3C 321: INTERACTING GALAXIES WITH A POWERFUL BLACK HOLE**
The system of intertwined galaxies 3C 321, a source of radio emission, lies at a distance of 1.2 billion light-years in the constellation Serpens. The incredible jet emanating from the galaxies, visible in this image in blue, is powered by a supermassive black hole. The distance between the galaxies themselves is 20,000 light-years, and the jet stretches farther into space than that.

Opposite **THE GALACTIC ROSE OF ARP 273**
Arp 273, designated so from Halton Arp's catalog of interacting galaxies, consists of a pair of gravitationally locked galaxies lying 300 million light-years away in the constellation Andromeda. The largest, elegantly armed spiral on top, UGC 1810, features a disk that has been distorted into the shape of a rose, tugged on by the gravitational pull of the edge-on galaxy below it, UGC 1813.

DANCING GALAXIES CREATE A NEW BURST OF STARS

The Hickson 31 Galaxy Group consists of four dwarf galaxies that lie 166 million light-years away, and give us a portrait of several galaxies whose interactions are producing new clusters of hot, blue stars.

Above **THE STRANGE, WARPED APPEARANCE OF THE INTEGRAL SIGN GALAXY**

UGC 3697, also known as the Integral Sign Galaxy, is a distorted object lying at a distance of 150 million light-years in the constellation Camelopardalis. Astronomers believe the unusual, scrolled edges of the galaxy resulted from interaction with a nearby dwarf galaxy.

Overleaf **HUBBLE VIEWS THE GALAXY'S CORE IN MINUTE DETAIL**

A color composite image of the Milky Way's center taken in infrared light reveals massive stars and swirls of hot, ionized gas in the region 300 light-years from the galaxy's core. This sharpest-ever image of the center of the galaxy reveals objects as small as 20 times the diameter of our solar system.

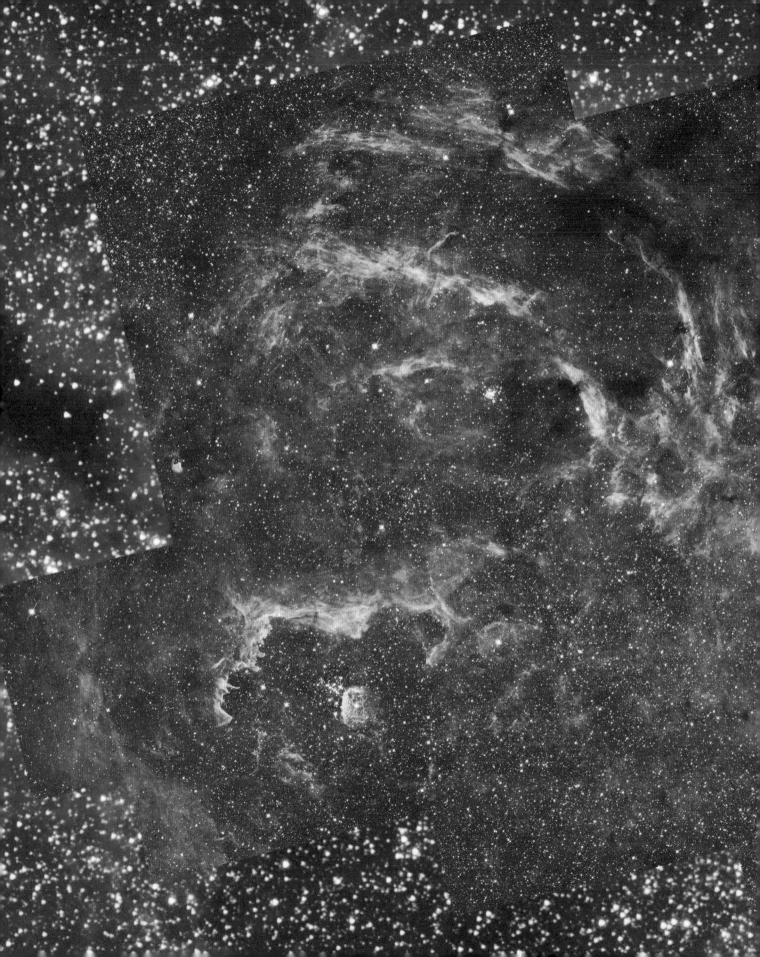

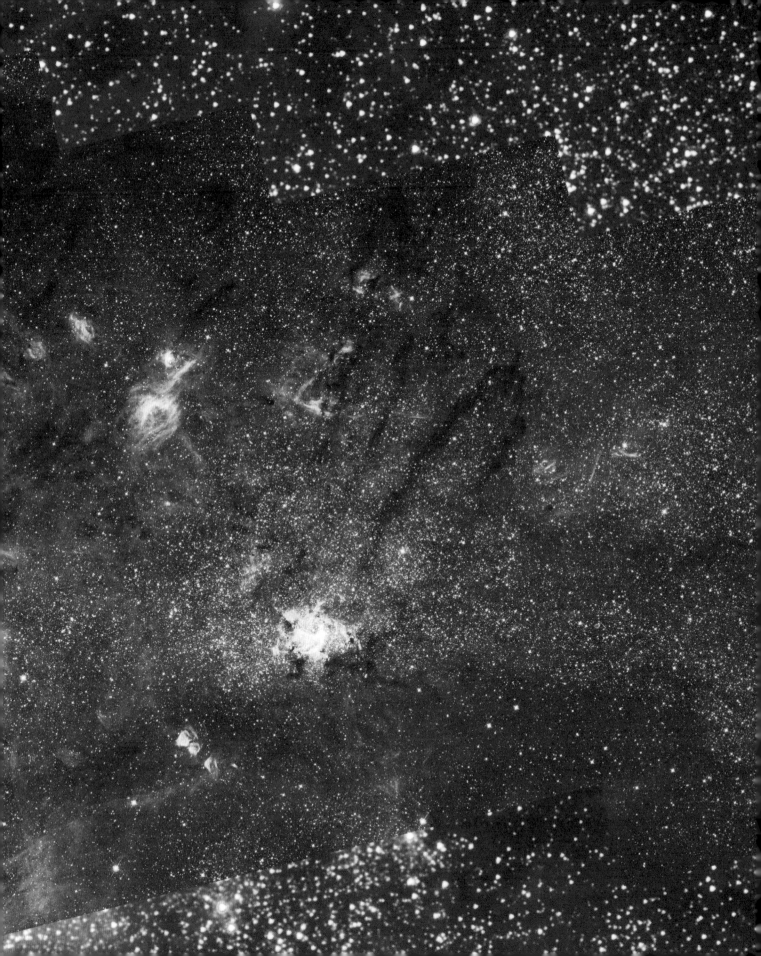

RESOURCES

Alfaro, Emilio J., Enrique Pérez, and José Franco, eds. *How Does the Galaxy Work? A Galactic Tertulia with Don Cox and Ron Reynolds.* Boston: Kluwer Academic Publishers, 2004.

Appenzeller, Immo. *High-Redshift Galaxies: Light from the Early Universe.* New York: Springer-Verlag, 2009.

Arp, Halton. *Catalogue of Discordant Redshift Associations.* Montreal: Apeiron Montreal, 2003.

———. *Quasars, Redshifts, and Controversies.* Berkeley, Calif.: Interstellar Media, 1987.

———. *Seeing Red: Redshifts, Cosmology, and Academic Science.* Montreal: Apeiron Montreal, 1998.

Combes, Françoise. *Mysteries of Galaxy Formation.* New York: Springer-Verlag, 2010.

Ferris, Timothy. *Galaxies.* New York: Stewart, Tabori, and Chang, 1982.

Hodge, Paul. *Atlas of the Andromeda Galaxy.* Seattle, Wash.: University of Washington Press, 1981.

———. *Galaxies.* Cambridge, Mass.: Harvard University Press, 1986.

Hubble, Edwin. *The Realm of the Nebulae.* New Haven, Conn.: Yale University Press, 2013.

Jones, Mark H., Robert J. A. Lambourne, and Stephen Serjeant, eds. *An Introduction to Galaxies and Cosmology.* Second ed. New York: Cambridge University Press, 2015.

Keel, William C. *The Road to Galaxy Formation.* New York: Springer-Verlag, 2002.

Mackie, Glen. *The Multiwavelength Atlas of Galaxies.* New York: Cambridge University Press, 2011.

Mulchaey, John S., Alan Dressler, and Augustus Oemler, eds. *Clusters of Galaxies: Probes of Cosmological Structure and Galaxy Evolution.* New York: Cambridge University Press, 2004.

Peterson, Bradley M. *An Introduction to Active Galactic Nuclei.* New York: Cambridge University Press, 1997.

Sandage, Allan, Mary Sandage, and Jerome Kristian, eds. *Galaxies and the Universe.* Chicago: University of Chicago Press, 1975.

Saviane, I., V. D. Ivanov, and J. Borissova, eds. *Groups of Galaxies in the Nearby Universe.* New York: Springer-Verlag, 2007.

Schneider, Peter. *Extragalactic Astronomy and Cosmology: An Introduction.* New York: Springer-Verlag, 2006.

Schultz, David. *The Andromeda Galaxy and the Rise of Modern Astronomy.* New York: Springer-Verlag, 2012.

Sheehan, William, and Christopher J. Conselice. *Galactic Encounters: Our Majestic and Evolving Star-System, From the Big Bang to Time's End.* New York: Springer-Verlag, 2015.

Sparke, Linda, and John S. Gallagher. *Galaxies in the Universe: An Introduction.* New York: Cambridge University Press, 2000.

Struck, Curtis. *Galaxy Collisions: Forging New Worlds from Cosmic Crashes.* New York: Springer-Verlag, 2011.

Waller, William H. *The Milky Way: An Insider's Guide.* Princeton, N.J.: Princeton University Press, 2013.

———. and Paul W. Hodge. *Galaxies and the Cosmic Frontier.* Cambridge, Mass.: Harvard University Press, 2003.

Wray, James D. *The Color Atlas of Galaxies.* New York: Cambridge University Press, 1988.

PHOTOGRAPH CREDITS

Pages 2–3: NASA, ESO, NAOJ, Giovanni Paglioli, R. Colombari, and R. Gendler

Pages 4–5: P. Horálek and ESO

Page 6: NASA and The Hubble Heritage Team (AURA/STScI)

Pages 10–11: Tony Hallas

Page 12: NASA, ESA, Z. Levay and R. van der Marel (STScI), T. Hallas, and A. Mellinger

Pages 16–17: Yuri Beletsky/Las Campanas Observatory/Carnegie Institution

Page 20: Images courtesy of the Carnegie Observatories/Cindy Hunt

Page 26: NASA, ESA, P. van Dokkum (Yale University), S. Patel (Leiden University), and 3D-HST Team

Page 35: SSRO–South (S. Mazlin, J. Harvey, D. Verschatse, and R. Gilbert) and K. Ivarsen (UNC/CTIO/PROMPT)

Pages 36–37: NASA, JPL–Caltech

Pages 38–39: NASA, ESA, J. Dalcanton, B. F. Williams, and L. C. Johnson (University of Washington), the PHAT Team, and R. Gendler

Pages 40–41: Tony Hallas

Page 42: Adam Block, Mount Lemmon SkyCenter, University of Arizona

Page 43: Don Goldman

Pages 44–45: NASA and The Hubble Heritage Team (STScI/AURA)

Page 46: ESA, NASA, and P. Anders (Göttingen University Galaxy Evolution Group)

Page 47: NASA and The Hubble Heritage Team (STScI/AURA)

Pages 48–49: NASA, ESA, and The Hubble Heritage Team (STScI/AURA)

Page 50: Adam Block, Mount Lemmon SkyCenter, University of Arizona

Page 51: Adam Block, Mount Lemmon SkyCenter, University of Arizona

Page 52: Hubble Legacy Archive, ESA, NASA/Roberto Colombari

Page 53 (top): Adam Block

Page 53 (bottom): NASA, ESA, and D. Maoz (Tel-Aviv University and Columbia University)

Pages 54–55: R. Jay Gabany

Page 56: Adam Block

Page 57: Hubble Legacy Archive, ESA, NASA, Martin Pugh

Page 58 (top and bottom): Adam Block

Page 59: Warren Keller

Pages 60–61: Adam Block

Page 62: ESO, INAF–VST, Omegacam

Page 63: Adam Block

Page 64: P-A. Duc (CEA, CFHT), Atlas 3-D Collaboration

Page 65: Subaru Telescope (NAOJ), Hubble Space Telescope, Robert Gendler

Page 66: Chart32 Team/Johannes Schedler

Page 67: Subaru Telecsope (NAOJ) and Robert Gendler

Page 68: NASA, ESA, A. Gal-Yam (Weizmann Institute)

Page 69: Hubble Legacy Archive, ESA, NASA, and Bill Snyder

Page 70: NASA and The Hubble Heritage Team (STScI/AURA)

Page 71: NASA, ESA, and The Hubble Heritage Team (STScI/AURA)–ESA/Hubble Collaboration, and W. Keel (University of Alabama)

Pages 72–73: NASA, ESA, and The Hubble Heritage Team (STScI/AURA)–ESA/Hubble Collaboration

Pages 88–89: Yuri Beletsky/Las Campanas Observatory/Carnegie Institution

Pages 90–91: P. Horalek/Eso

Page 92: Yuri Beletsky/Las Campanas Observatory/Carnegie Institution

Page 93: Eso

Pages 94–95: NASA, SWIFT, S. Immler (Goddard), and M. Sigel (Penn State); Axel Mellinger (CMU)

Page 96: Bernhard Hubl

Page 97: ESO/VISTA VMC

Pages 98–99: Jason Jennings

Pages 100–5: NASA, ESA, Z. Levay and R. van der Marel (STScI), T. Hallas, and A. Mellinger

Pages 106–7: Adam Block

Page 126: Adam Block, Mount Lemmon SkyCenter, University of Arizona

Page 127: Subaru Telescope (NAOJ), Hubble Space Telescope, Kpno, Noao, Digitized Sky Survey, Spitzer Space Telescope, and R. Gendler

Page 128: Bernhard Hubl

Page 129: Adam Block

Pages 130–31: Subaru Telescope (NAOJ), Hubble Space Telescope, Local Group Galaxy Survey (Phil Massey, Pi), Mayall 4-Meter Telescope, R. Gendler

Page 132: NASA and T. M. Brown, C. W. Bowers, R. A. Kimble, A. V. Sweigart (NASA GSFC), and H. C. Ferguson (STScI)

Pages 133 and 134: Mark Hanson

Page 135 (top left): Chris Schur

Page 135 (top right): NASA, ESA, G. Kriss (STScI), and J. De Plaa (Sron Netherlands Institute for Space Research)

Page 135 (bottom left): NASA and The Hubble Heritage Team (STScI/AURA)

Page 135 (bottom right): NASA/CXC/SAO, NASA/STScI, NSF/NRAO/AUI/VLA

Pages 136–37: NASA, ESA, S. Baum and C. O'Dea (RIT), R. Perley and W. Cotton (NRAO/AUI/NSF), and The Hubble Heritage Team (STScI/AURA)

Page 138: NASA, A. S. Wilson (University of Maryland), P. L. Shopbell (Caltech), C. Simpson (Subaru Telescope), T. Storchi-Bergmann and F. K. B. Barbosa (UFRGS), and M. J. Ward (University of Leicester)

Page 139: NASA/CXC/MIT/C. Canizares, D. Evans, et. al., NSF/NRAO/VLA

Pages 140–41: NASA, ESO, NAOJ, Giovanni Paglioli, and R. Colombari

Page 142: Adam Block

Page 143: NASA/CXC/Digitized Sky Survey

Pages 144–45: Adam Block

Page 146: NASA/CXC/University of Wisconsin/Y. Bai, et. al.

Page 147: NASA/UMASS/D. Wang, et. al.

Pages 148–49: Susan Stoloby (SSC/CALTECH), et. al./JPL–CALTECH/NASA

Pages 172–73: NASA, ESA, and The Hubble Heritage Team (STScI/AURA)

Pages 174–75: Adam Block

Page 176: R. Jay Gabany

Page 177: NASA/CXC/SAO/D. Patnaude, et. al./ESO/VLT/NASA/JPL/CALTECH

Pages 178–79: Gemini Observatory, Aura, Travis Rector (University of Alaska–Anchorage)

Page 180: Jack Newton

Page 181 (top and bottom): Don Goldman

Pages 182–83: Tony Hallas

Page 184: Gerald Rhemann

Page 185: Adam Block

Page 186: E. Peng and H. Ford (JHU)/K. Freeman (ANU)/R. White (STScI)/CTIO/NOAO/NSF

Page 187: NASA/CXC/Wesleyan/R. Kilgard, et. al.

Page 188: NASA, ESA, and The Hubble Heritage Team (STScI/AURA)

Pages 189 and 190–91: Adam Block

Page 192: NASA/CXC/CfA/R. Kraft, et. al./MPIfR/ESO/APEX/A. Weiss, et. al./ESO/WFI

Page 193: The Hubble Heritage Team (AURA/STScI/NASA)

Pages 194–95: Hubble Legacy Archive/ESA/NASA

Page 196: Adam Block

Pages 197, 198–99: Tony Hallas

Page 200: Hubble Legacy Archive/NASA/ESA

Page 201: Adam Block

Page 202: Hubble Legacy Archive, NASA, ESA, José Jiménez Priego

Page 203: NASA, ESA, and The Hubble Heritage Team (STScI)

Page 204: Tony Hallas

Page 205: NASA, ESA, and T. Brown (STScI)

Page 206: Tony Hallas

Page 207: NASA, ESA, and The Hubble Heritage Team (STScI/AURA)

Pages 208–9: SSRO–SOUTH/J. Harvey/S. Mazlin/D. Verschatse/J. Joaquin Perez (UNC/CTIO/PROMPT)

Page 230: Adam Block

Page 231: NASA, ESA, and S. Farrell (Sydney Institute for Astronomy, University of Sydney)

Page 232 (top): NASA, ESA, M. Postman (STScI), T. Lauer (NOAO), and The Clash Team

Page 232 (bottom): NASA, ESA, and The Hubble Heritage Team

Page 233: NASA, ESA, and A. Nota (STScI)

Page 234: NASA, ESA, and The Hubble Heritage Team (STScI/AURA)

Page 235 (top): NASA and Michael Corbin (CSC/STScI)

Page 235 (bottom): W. Keel (University of Alabama), M. Ledlow (Gemini Observatory), F. Owen (NRAO), AUI, NSF, NASA

Page 236: NASA, ESA, and The Hubble Heritage Team (STScI/AURA)–ESA/Hubble Collaboration

Page 237: NASA, ESA, and The Hubble Heritage Team (STScI/AURA)–ESA/Hubble Collaboration

Page 238: NASA/CXC/CfA/D. Evans, et. al., NASA/STScI/NSF/VLA/CfA/D. Evans, et. al., STFC/JBO/MERLIN

Page 239: NASA, ESA, and The Hubble Heritage Team (STScI/AURA)

Page 240: NASA, ESA, J. English (U. Manitoba)

Page 241: Don Goldman

Pages 242–43: NASA, ESA, and Q. D. Wang (University of Massachusetts, Amherst)

Page 250: FORS, 8.2-METER VLT ANTU, ESO

ACKNOWLEDGMENTS

As with any book project, many people have contributed their talents and guidance to my work, beyond the writing and compiling. That said, any shortcomings of this book are wholly my own. But I want to acknowledge a number of generous people who have contributed to make this book happen. They start with my family, Lynda Eicher and Chris Eicher, who as always have supported this project from the outset. My superb editors at Clarkson Potter, Angelin Borsics and Jenni Zellner, helped shape the book from day one, and the rest of the team worked hard to ensure its sucess: designer Mia Johnson, illustrator Irene Laschi, production editor Joyce Wong, and production manager Phil Leung. My agent, Jennifer Weltz, has been a huge source of advice and suggestions, as was the original agent at my firm, Laura Biagi, before she departed for other adventures.

Great thanks are due to one of the world's leading experts on galaxies, Jay Gallagher of the University of Wisconsin, for kindly contributing the book's foreword. Jay's knowledge of galaxy research stretches far back beyond my first days of knowing him, back in the 1980s, when he was at Lowell Observatory.

I want to acknowledge the generous help of several folks at Kalmbach Media, the publishers of *Astronomy* magazine. Michael Bakich helped to sort and find images through a mountain of fantastic shots he receives from amateur astronomers all across the world. Thanks are also due to Steve George and Becky Lang, who kindly allowed use of some diagrams that first appeared in *Astronomy* and *Discover* magazines.

I also want to thank several friends for general encouragement, contributing advice and expertise, and helping me along through a period of simultaneous projects. They include Richard Dawkins, Garik Israelian, Brian May, Robin Rees, Brian Skiff, and Glenn Smith. I thank Timothy Ferris for his classic book *Galaxies*, published in 1980, that inspired me to want to write about the subject "someday."

Thanks are due to the very generous Cynthia Hunt of Carnegie Observatories for sending the original photographs of the Andromeda Galaxy made by Edwin Hubble in 1923.

And lastly, but far from least, I want to thank the generous photographers who enabled me to use their imagery in this book. The quality of galaxy photographs taken by backyard astronomers has skyrocketed in the last decade, and I am proud to be able to include their efforts here. These heroic imagers are Adam Block, Ken Crawford, Thomas V. Davis, Bob Fera, R. Jay GaBany, Don Goldman, Dietmar Hager, Tony Hallas, Mark Hanson, Bernhard Hubl, Jason Jennings, Warren Keller, Jack Newton, Gerald Rhemann, and Chris Schur.

THE GRAND SPIRAL FACE-ON GALAXY NGC 1232
One of the gems of the southern sky, NGC 1232
is a face-on many-armed spiral hosted by the
constellation Eridanus. Its intricate armed structure
shows a gravitational swirl of interstellar gas, star
clusters, and star forming regions. The galaxy lies
60 million light-years away.

INDEX